Amplifier Applications
of Op Amps

Other Books by Jerald G. Graeme

Optimizing Op Amp Performance
Photodiode Amplifiers: Op Amp Solutions

Amplifier Applications of Op Amps

Jerald G. Graeme
Principal Engineer
Gain Technology Corporation

McGraw-Hill

New York San Francisco Washington, D.C. Auckland Bogotá
Caracas Lisbon London Madrid Mexico City Milan
Montreal New Delhi San Juan Singapore
Sydney Tokyo Toronto

McGraw-Hill

A Division of The McGraw-Hill Companies

2 3 4 5 6 7 8 9 0   AGM/AGM   9 0 4 3 2 1 0

P/N 0-07-134643-0
PART OF
ISBN 0-07-134642-2

The sponsoring editor for this book was Scott Grillo, the editing supervisor was Peggy Lamb, and the production supervisor was Sherri Souffrance. This book was set in Century Schoolbook by Ron Painter of McGraw-Hill's Professional Book Group composition unit. Illustrated by Lola E. Graeme.

Printed and bound by Quebecor Martinsburg.

This book is printed on recycled, acid-free paper containing a minimum of 50% recycled, de-inked fiber.

Contents

Preface

Op amps serve as the fundamental element for widely varied amplifier functions. In the most basic applications, they provide voltage gain through noninverting, inverting, and difference amplifier configurations. Variations on these basic configurations realize gain options through summing, variable-gain, and switched-gain amplifiers. Introducing positive feedback extends the op amp's function to current-mode amplifiers and later to transconductance amplifiers, which convert voltage to current. In companion to the latter application, transimpedance amplifiers convert current to voltage. Another variation encloses two op amps in a common feedback loop to form composite voltage amplifiers that improve accuracy and speed. Next, combining two noninverting amplifiers produces differential-output amplifiers that extend the benefit of common-mode rejection to the output side of a signal transmission. Instrumentation amplifiers serve as the differential line receivers for this transmission with three configurations that provide high performance, lower cost, or differential outputs. Bridge amplifiers also provide differential monitoring in a variety of configurations that counteract the inherent nonlinearity of the basic bridge response. Clamping amplifiers limit output voltage swing through numerous configurations having differing degrees of precision and control. Phantom amplifiers reduce the wiring requirements of multichannel remote monitoring by coupling a third differential signal through the common-mode signals of two other differential channels.

The noninverting, inverting, and difference amplifier connections represent the op amp's basic voltage-amplifier configurations. While the inverting and noninverting connections remain similar, they display differences beyond that of the obvious gain polarity and input impedance distinctions. Feedback analysis reveals further distinctions in common-mode error and bandwidth. For higher-gain applications of these configurations, introducing tee feedback reduces dc offset error through an equivalent feedback resistance that more accurately matches a traditional offset compensation resistor. The difference-amplifier connection combines elements of the noninverting and inverting connections to produce signal subtraction as well as rejection of common-mode input signals. Extensions to the basic difference amplifier accommodate high-level common-mode signals or permit simple adjustment of common-mode rejection.

For each of these basic amplifier configurations, variations on the circuit's feedback control produce gain options in the form of signal summation, variable gain, or switched gain. Signal summation provides weighted gains to a number of input signals, producing a composite output signal. While normally associated with the inverting op amp configuration, signal summation extends to the noninverting and difference-amplifier configurations as well. Variable or switched gains result from potentiometers, analog multipliers, or switches connected in the op amp's feedback network. In the simplest case, just replacing one of the circuit's traditional feedback resistors with a potentiometer produces continuously variable gain for both the inverting and the noninverting configurations. As a first alternative to the potentiometer, connecting an analog multiplier in the feedback path produces a variable effective resistance for continuously variable electronic control of gain. As a second alternative, connecting feedback switches that access multiple feedback configurations also provides electronic control, but in discrete gain steps.

While universally known for processing signal voltages, the op amp also performs analogous tasks with signal currents. The equivalent current amplifiers accept an input signal current, process it, and deliver an output current signal. The straightforward processing of such signals would first convert the signal current to a voltage, process the intermediate result in voltage mode with conventional circuitry, and then reconvert the resulting voltage to an output current. However, retaining the signal in the current mode avoids the added complexity and errors introduced by these conversions. Single op amp circuits provide this processing with performance characteristics that parallel those of voltage-mode counterparts. To do so, these circuits combine negative and positive feedback in producing a current output in response to a current input. Examples presented demonstrate this feedback control for noninverting, inverting, difference, integrating, and differentiating current amplifiers.

Extending the departure from voltage-based processing, transimpedance and transconductance amplifiers translate between voltage and current signals. Transimpedance amplifiers convert a current input into a voltage output and transconductance amplifiers provide the converse role, converting an input voltage into an output current. One op

amp configuration emerges as the de facto standard for the transimpedance amplifier role due to its circuit simplicity and the near ideal
terminal impedances it presents. The current-to-voltage configuration
consists of just an op amp and a feedback resistor and presents near
zero impedances at both its input and its output terminals. A second op
amp configuration serves as the basis for transconductance amplifiers
through its ability to implement the positive feedback needed for the
voltage-to-current conversion. In this configuration, a modified difference amplifier includes feedback networks connected to both inputs of
an op amp. Analysis reduces the implementation of this feedback configuration and its derivatives to simple design equations.

Most op amp applications enclose only one amplifier within a given
feedback loop but composite amplifiers enclose two. The two amplifiers connect in series, producing increased gain, accuracy, and bandwidth. This connection boosts the circuit's open-loop gain to the product of the individual amplifier gains, which may be either open-loop
or closed-loop. In either case, the increased open-loop gain of the composite connection reduces the error contributions of both op amps and
extends the circuit bandwidth. However, the two amplifiers in the
same loop produce a two-pole composite response, requiring phase
compensation. Six phase compensation methods address this requirement. Five of these follow traditional phase compensation philosophy,
modifying the A_{OL} response, and the sixth modifies the $1/\beta$ feedback
response instead. A one-time analysis of the six compensation methods yields standard design equations that replace the application-
specific analyses otherwise required.

Differential-output amplifiers extend the benefit of common-mode
rejection to the output side of a signal transmission. Widely used differential-input amplifiers provide this benefit at the input side for
rejection of coupled noise. Extending differential operation to the
amplifier's output permits a similar rejection there. For this purpose,
the differential noninverting amplifier develops a differential output
signal in response to a differential or single-ended input signal. In the
process, this amplifier retains the rejection of common-mode input
signals, increases output voltage swing, and frequently improves the
slew rate. The basic form of this circuit prevents direct common-mode
corruption of the differential signal, but it still transfers any common-
mode input signal directly to the circuit output. There, line impedance imbalances potentially mix the differential and common-mode
signals. Adding common-mode feedback through a third op amp prevents this corruption by removing the common-mode signal from the
output lines. In either configuration, the differential noninverting
amplifier also performs the task of single-ended-to-differential con-

version, as often required for signal transmission on a two-wire pair.

Differential signal transmission and common-mode rejection of coupled errors characterize the role of the instrumentation amplifier. While sharing the differential-input nature of an op amp, the instrumentation amplifier includes committed feedback that retains high impedance at both amplifier inputs for superior immunity to line-induced errors. Analyses develop the gain, bandwidth, offset, and common-mode rejection responses of three instrumentation amplifier configurations. First, the three-op-amp configuration generally provides the highest overall performance, using two op amps in an input circuit that extracts the differential signal, followed by a difference amplifier that removes the common-mode signal. Next, two op amps form a lesser used but simpler instrumentation amplifier with an application-specific benefit. Compared with the three-op-amp solution, this configuration restricts gain range, bandwidth, and signal swing but extends high-frequency CMRR. Finally, adding a second difference amplifier to the three-op-amp configuration produces a differential-output instrumentation amplifier for the continuance of differential signal transmission.

Bridge amplifiers address the particularly demanding requirements imposed by the low-level outputs typical of transducers. Connecting the transducer in a bridge configuration greatly eases these requirements by reducing the signal monitored to its deviation from the quiescent state. However, the bridge connection also introduces a common-mode signal, requiring differential measurement of the transducer response. Instrumentation amplifiers most commonly serve this purpose, providing common-mode rejection of this extraneous signal. Alternately, op amps replace the instrumentation amplifier, using subtraction to remove the common-mode signal. In either case, the bridge connection of a single transducer introduces nonlinearity to the circuit response. A variety of two-amplifier bridge circuits produce linearized, high-gain responses through bias or feedback control of the bridge. To linearize the response, each of these circuits eliminates the transducer's influence on its own bias current.

Clamping amplifiers or feedback limiters provide amplitude control for signal clipping, signal squaring, and overload protection. To define the clamping voltage level, a zener diode or its bandgap equivalent generally offer the simplest solutions. Such solutions provide a fixed higher-voltage clamping action of moderate precision. Adding op amps and diode bridges to the clamping circuit improves precision by reducing the output impedance, sharpening the clamping transition, balancing the clamping symmetry, and increasing speed. Replacing the zener diodes with power-supply-referenced voltage dividers and rectifying junctions expands clamping options to a greater range of voltages. Adding feedback control over the rectifying junctions sharpens and flattens the clamping response while extending the clamping range and accuracy to very low levels. Finally, replacing the clamping voltage reference with a control signal permits electronic variation of clamping levels.

Phantom amplifiers continue the wiring reduction benefit of the venerable phantom circuit and extend its frequency response down to DC. These amplifiers instrument the differential measurement of remote signals that inherently acquire common-mode errors and potentially develop crosstalk through shared common returns. With these amplifiers, differential measurement greatly improves remote monitoring accuracy through common-mode rejection of coupled noise and ground potential differences. This measurement also removes the crosstalk effects of remote multichannel instrumentation, when dedicated two-wire pairs serve each monitor channel. Phantom circuits and phantom amplifiers reduce this wiring requirement by one-third. They do so by coupling a third differential signal as the difference between the common-mode signals of two other monitor channels. Differential measurement with the transformer-coupled phantom circuit becomes impractical at lower frequencies due to the physical size of the transformers then required. Replacing the transformers with differential amplifiers forms phantom amplifiers that provide the same wiring efficiency but without the low-frequency limitation.

I wish to thank the many op amp users with whom I have discussed application requirements over the years. Their inquiries led to the investigation and development of much of the material presented here. My thanks to *EDN* and *Electronic Design* for publishing the articles that initiated many of these inquiries. Finally, I wish to thank my wife Lola for her accurate and attractive rendering of the illustrations and for the rewarding feeling of mutual involvement in preparing this book.

Jerald G. Graeme

Basic Voltage Amplifiers

The noninverting, inverting, and difference amplifier connections represent the op amp's fundamental voltage amplifier configurations. While the inverting and noninverting connections remain similar, they display differences beyond that of the obvious gain polarity and input impedance distinctions. Feedback analysis reveals significant distinctions in common-mode error and bandwidth as well. For higher-gain applications of these configurations, introducing tee feedback reduces dc offset error through an equivalent feedback resistance. There, the tee replaces a high-value feedback resistor with smaller resistances having better resistance accuracy. This replacement permits more accurate cancellation of the offset introduced by the amplifier's input bias currents. The difference amplifier connection combines elements of the noninverting and inverting connections to produce signal subtraction that rejects common-mode input signals. Analysis defines this rejection in terms of resistor matching errors and the op amp's inherent common-mode rejection error. Extensions to the basic difference amplifier accommodate high-voltage common-mode signals or permit simple adjustment of common-mode rejection.

1.1 Single-Input Amplifiers

The noninverting and inverting connections of an op amp form the most basic voltage amplifiers of the class, each having a single signal input. Comparing the two, the noninverting amplifier provides greater bandwidth and input impedance but also introduces a common-mode error signal. Also, the noninverting configuration presents a fundamentally capacitive rather than resistive input impedance. Alternately, the inverting configuration presents a resistive but considerably lower input impedance.

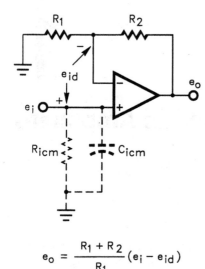

$$e_o = \frac{R_1 + R_2}{R_1}(e_i - e_{id})$$

Figure 1.1 Noninverting op amp configuration presents a high input impedance, dominated by C_{icm}, and amplifies both signal e_i and error e_{id} by the same voltage gain.

1.1.1 Noninverting amplifiers

Shown in Fig. 1.1, the basic noninverting configuration generally provides better overall accuracy than the inverting configuration described next. The noninverting configuration develops the familiar voltage gain of $G_n = 1 + R_2/R_1$, and the circuit amplifies both the applied input signal e_i and the op amp's input error signal e_{id} by this same gain. This produces $e_o = G_n(e_i - e_{id})$, where e_{id} includes all of the input-referred errors of the op amp. Basic input errors result from the op amp's input offset voltage, input bias currents, and input noise voltage. In addition, the op amp's finite values of open-loop gain, common-mode rejection, and power-supply rejection introduce signal-related input errors. As such, the G_n amplification of e_{id} presents a fairly extensive indication of the noninverting circuit's output errors. A detailed analysis[1] of e_{id} defines it as

$$e_{id} = V_{OS} - I_{B+}R_{S^+} + I_{B-}R_{S^-} + e_n + \frac{e_o}{A} + \frac{e_{icm}}{\text{CMRR}} + \frac{\delta V_S}{\text{PSRR}}$$

where R_{S^+} and R_{S^-} represent the net source resistances presented to the op amp's noninverting and inverting inputs by the applied signal source and the feedback network. The V_{OS}, I_B, and e_n error terms represent obvious input-referred error sources, and the referenced analysis develops the error terms corresponding to open-loop gain A, CMRR, and PSRR.

The noninverting configuration presents a high input impedance to the signal source but parasitic capacitance limits this impedance at higher frequencies. As shown in the figure, the op amp presents its common-mode input resistance and capacitance R_{icm} and C_{icm} at the e_i input. While R_{icm} represents a very high resistance, C_{icm} bypasses this and controls the net input impedance over much of the op amp's useful frequency range. For bipolar-input op amps, C_{icm} typically takes control of the input impedance at around 1 kHz. For FET input op amps, the extremely high value of R_{icm} makes the transition occur at a very low frequency, and R_{icm} can be ignored for most applications.

Analysis of the noninverting circuit's error model reveals a unique correspondence between its ideal gain G_n and the circuit's feedback factor β. Shown in Fig. 1.2, that model represents the circuit's feedback network by a β transmission block that couples the signal βe_o to the amplifier's inverting input. Thus the feedback factor β equals the fraction of the output fed back to that input. Analysis of the model defines an output signal of $e_o = (1/\beta)(e_i - e_{id})$, making $G_n = 1/\beta$, and this gain correspondence holds only for the noninverting configuration. Further evaluation of $G_n = 1/\beta$ defines the circuit's feedback factor as $\beta = 1/G_n = R_1/(R_1 + R_2)$, and this factor equals the voltage divider ratio of the feedback network, as seen from the circuit output.

A prior publication[2] presents a more detailed development of this

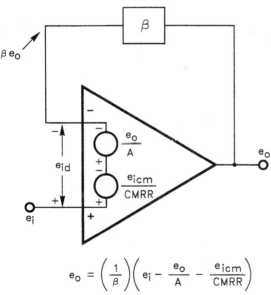

$$e_o = \left(\frac{1}{\beta}\right)\left(e_i - \frac{e_o}{A} - \frac{e_{icm}}{CMRR}\right)$$

Figure 1.2 Modeling the noninverting configuration with just two of the e_{id} input error sources simplifies a later comparison with the noninverting configuration.

gain and feedback relationship along with other performance relationships to β. However, for the basic comparison of noninverting and inverting configurations here, simplified error and bandwidth analyses suffice. As described, the noninverting circuit amplifies the op amp's input error signal e_{id} by a gain of $1/\beta$. In the model in Fig. 1.2 two fundamental error signals form a simplified $e_{id} = e_o/A + e_{icm}/\mathrm{CMRR}$, and they result from the finite levels of the op amp's open-loop gain and common-mode rejection. These two error components permit a fundamental performance comparison without including the detail of the other e_{id} components. In the model, finite open-loop gain requires the e_o/A error signal between the op amp inputs to develop the e_o output signal from the amplification provided by gain A. Finite CMRR and the presence of a signal at the op amp's noninverting input produce the e_{icm}/CMRR error signal. Signal e_i drives this input, making $e_{icm} = e_i$. Substituting $e_{id} = e_o/A + e_{icm}/\mathrm{CMRR}$ and $e_{icm} = e_i$ in $e_o = 1/\beta(e_i - e_{id})$ and solving defines the circuit's fundamental response as

$$e_o = \left(\frac{1}{\beta + 1/A} \right)\left(\frac{\mathrm{CMRR} - 1}{\mathrm{CMRR}} \right) e_i$$

As expected, this response deviates from the ideal $1/\beta$ gain due to the finite values of A and CMRR.

At higher frequencies, both A and CMRR decline, producing the response roll off that defines bandwidth. Fortunately, CMRR appears in both the numerator and the denominator of this equation, and this mitigates its effect. Also, the typical high-frequency CMRR generally exceeds the high-frequency value of open-loop gain A. Combined, these two CMRR characteristics generally make the associated bandwidth limitation secondary to that of the gain roll off. Then, the 3-dB bandwidth limit occurs where A drops to the level of the feedback demand for gain, as illustrated in Fig. 1.3. There, the open-loop gain response supports the feedback demand imposed by the circuit's $1/\beta$ curve up to their intercept. Beyond this intercept, the op amp lacks the gain required to support the circuit's ideal response and the actual response rolls off, producing the bandwidth limit. The intercept frequency f_i defines the circuit's -3-dB bandwidth at the coincident point of the A and $1/\beta$ curves.[3] At this intercept, the open-loop gain responses of most op amps follow a single-pole roll off, making $A \approx f_c/f$, where f_c is the unity-gain crossover frequency of the op amp. Then at the intercept, $A = f_c/f_i = 1/\beta$ and $f_i = \beta f_c = \mathrm{BW}_n$ defines the intercept frequency and bandwidth. From before, $G_n = 1/\beta$, making $\mathrm{BW}_n = \beta f_c = f_c/G_n$, and rearranging this expression reflects that the noninverting configuration realizes the full gain–bandwidth product of the op amp, $\mathrm{GBW}_n = f_c$.

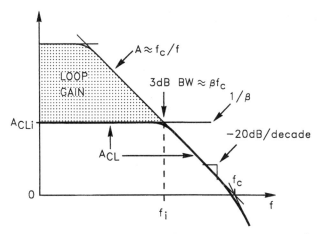

Figure 1.3 Bode plot analysis of the noninverting case reveals the classic bandwidth limit of an op amp circuit at $BW = f_i \approx \beta f_c$.

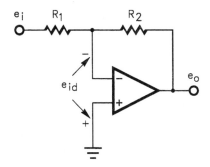

$$e_o = -\frac{R_2}{R_1} e_i - \frac{R_1 + R_2}{R_1} e_{id}$$

Figure 1.4 Compared with the noninverting configuration, the inverting configuration continues to amplify the e_{id} error with the same voltage gain but amplifies the e_i signal with a reduced gain.

1.1.2 Inverting amplifiers

Compared with their noninverting counterparts, inverting amplifiers remove common-mode error signals but increase other errors, reduce input impedance, and reduce the gain–bandwidth product. Shown in Fig. 1.4, the basic configuration presents a resistive input impedance equal to R_1. The circuit develops the familiar signal gain $G_i = -R_2/R_1$ but continues to amplify the op amp's input error signal e_{id} by the higher gain $G_e = G_n = 1 + R_2/R_1$. Using superposition and grounding the e_i input shows that the circuit remains a noninverting amplifier for signal e_{id} and produces this higher error-signal gain. Then $e_o =$

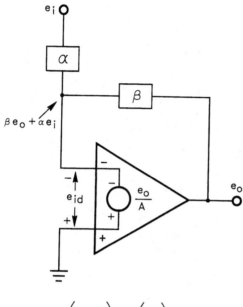

$$e_o = \left(1 - \frac{1}{\beta}\right)e_i - \left(\frac{1}{\beta}\right)\frac{e_o}{A}$$

Figure 1.5 Error model for the inverting configuration eliminates the common-mode error source of the previous noninverting case and adds an α feedforward transmission block.

$G_i e_i - G_e e_{id}$, and analysis of an error model confirms this differentiation of signal and error gains.

Shown in Fig. 1.5, that model repeats the β feedback transmission block of the noninverting case and adds an α input transmission block. The latter block accounts for the indirect drive of the op amp input by the e_i signal. Previously, that signal drove an op amp input through a direct connection, but here the feedback network attenuates the signal coupled from e_i to the amplifier. For this signal, the voltage divider action of the network reduces the portion of e_i coupled to the op amp's inverting input by the factor $\alpha = R_2/(R_1 + R_2)$. As before, this network also attenuates the signal coupled there by e_o through the feedback factor $\beta = R_1/(R_1 + R_2)$. Using these α and β factors, an analysis of the model defines an output signal $e_o = -(\alpha/\beta)e_i - (1/\beta)e_{id} = (1 - 1/\beta)e_i - (1/\beta)e_{id}$. From this equation, $|G_i| = |1 - 1/\beta| < |G_e| = |1/\beta|$. Thus the inverting configuration amplifies the error signal by a greater gain magnitude than that supplied to the intended input signal.

Although the inverting configuration makes $|G_e| > |G_i|$, this configuration does reduce the error signal amplified by G_e. As modeled, this circuit supports no common-mode input signal, removing the e_{icm}/CMRR input error signal of the previous noninverting case. This reduces the e_{id} input error signal to simply $e_{id} = e_o/A$. Substituting this in $e_o = (1 - 1/\beta)e_i - (1/\beta)e_{id}$ and solving for e_o defines the circuit's fundamental response as

$$e_o = \left(1 - \frac{1}{\beta}\right)\left(\frac{1}{1 + 1/A\beta}\right)e_i$$

As expected, this response deviates from the ideal $(1-1/\beta)$ gain due to the finite value of the open-loop gain A.

At higher frequencies, A declines, again producing a response roll off that defines the circuit's bandwidth. As with the noninverting case described earlier, the inverting circuit's feedback factor and open-loop gain response define the -3-dB bandwidth limit. For a given feedback network the gain magnitudes of the noninverting and inverting cases differ but the feedback factor remains the same. In both cases, $\beta = R_1/(R_1 + R_2)$ and $\text{BW} = \beta f_c$, where f_c is the unity-gain crossover frequency of the op amp. Thus for the inverting configuration, $\text{BW}_i = \beta f_c = f_c/(G_i - 1)$. This makes the inverting circuit's gain–bandwidth product $\text{GBW}_i = G_i f_c/(G_i - 1)$, reflecting that, unlike the noninverting configuration, the inverting configuration does not realize the full gain–bandwidth product $\text{GBW} = f_c$ of the op amp. Compared with the noninverting configuration, the inverting configuration reduces GBW by a fraction equal to $G_i/(G_i - 1)$.

1.2   Offset Compensation and Tee Networks

Higher-gain applications of the basic voltage amplifiers become vulnerable to a practical limit imposed by dc offset. In such cases, the associated higher R_2 feedback resistances produce dominant offset errors in conjunction with the op amp's input bias current. Adding a compensation resistor routinely limits this offset effect by making the op amp's two input currents produce canceling effects. However, this compensation requires resistance matching, which the poor tolerance of high-valued resistors compromises. Then, converting the conventional R_2 feedback resistor to a tee network reduces the resistance values required to improve the offset compensation.

1.2.1   Offset compensation

First consider the offset compensation technique as illustrated in Fig. 1.6. This figure shows the basic inverting amplifier with the commonly added compensation resistor R_C. Adding R_C introduces an offset voltage with I_{B^+}, which counteracts that produced by R_2 and I_{B^-}. The op amp draws the dc bias current for its inverting input I_{B^-} from the feedback network and produces a first component of offset error. That current would split between the R_1 and R_2 resistors of the network except for the circuit's feedback control. Feedback forces the op amp's

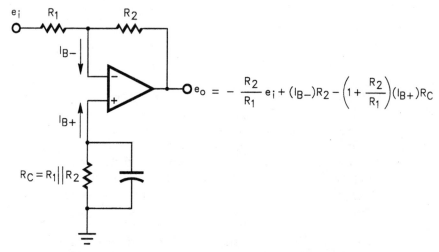

Figure 1.6 Adding an R_C compensation resistor reduces the output offset error developed by amplifier input current through a counteracting offset component.

two inputs to the same voltage, preventing I_{B-} from developing a voltage upon R_1. To do so, the feedback absorbs I_{B-} through R_2, developing an output offset component equal to $I_{B-}R_2$. The large R_2 resistance values commonly used for high gain levels make this a major offset component, and this component exhibits a large thermal drift due to the high temperature coefficient of I_{B-}.

To counteract this offset, adding the R_C resistor in series with the op amp's noninverting input produces a compensating effect, and capacitive bypass of this resistor largely removes its noise effect. The reaction of R_C with the op amp's noninverting input current produces a dc offset voltage of $-(I_{B+})R_C$ at the op amp's noninverting input. To this voltage, the circuit appears as a noninverting amplifier that supplies a gain of $1 + R_2/R_1$ and produces an output offset component equal to $-(1 + R_2/R_1)(I_{B+})R_C$. This component counteracts that produced by I_{B-} to the degree permitted by circuit mismatches. Matching errors between the amplifier input currents and between resistances retain a reduced output offset error of $V_{OSO} = (I_{B-})R_2 - (1 + R_2/R_1)(I_{B+})R_C$. Selecting $R_C = R_1 \| R_2 = R_1R_2/(R_1 + R_2)$ and ideal resistor matching reduces this to $V_{OSO} = (I_{B-} - I_{B+})R_2$. Then, ideal bias current matching makes $I_{B-} = I_{B+}$ and $V_{OSO} = 0$. However, practical mismatches limit this offset reduction to about a factor of 10 improvement over the initial offset and drift produced by the amplifier's input current. Large values of R_2 introduce greater resistance tolerance errors, which further compromise this offset compensation for both the inverting and the noninverting cases.

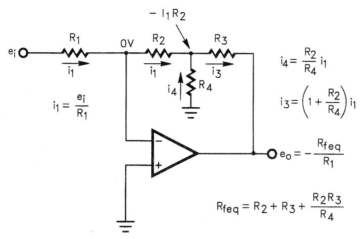

Figure 1.7 For high-gain applications, replacing the conventional R_2 feedback resistor with a tee network produces a very high equivalent resistance with lower resistance values having better tolerance accuracies.

1.2.2 Inverting amplifiers with tee feedback

For high-gain applications, replacing R_2 with a resistor tee reduces offset error through reduced resistance levels. The reduced resistances improve resistor tolerances to increase the probability of the resistance matching required for the offset compensation. However, the tee network also imposes a limit to this offset reduction by amplifying the op amp's input offset voltage. This compromise serves as a guide in the design of the tee alternative for a given application.

For the inverting amplifier, Fig. 1.7 shows the replacement of the original R_2 with a tee that produces an equivalent feedback resistance using smaller resistance levels. Considering signal conditions first, the circuit's feedback action accepts the input current $i_1 = e_i/R_1$ at the amplifier's inverting input. In doing so, the feedback signal develops a voltage of $-i_1 R_2$ across R_2. For the basic inverting amplifier, this would be the circuit's output signal. However, the inclusion of the voltage divider formed by R_3 and R_4 here increases the output voltage e_o required to develop this signal. The $-i_1 R_2$ voltage also drops across R_4, where it develops $i_4 = i_1 R_2/R_4$, and that current combines with i_1 to form $i_3 = i_1 + i_4 = (1 + R_2/R_4)i_1$. The flow of i_3 in R_3 develops a voltage that combines with the original $-i_1 R_2$ to develop an output voltage of $e_o = -i_1 R_2 - i_3 R_3 = -[R_2 + (1 + R_2/R_4)R_3]i_1$, or

$$e_o = -\frac{R_2 + R_3 + R_2 R_3/R_4}{R_1} e_i$$

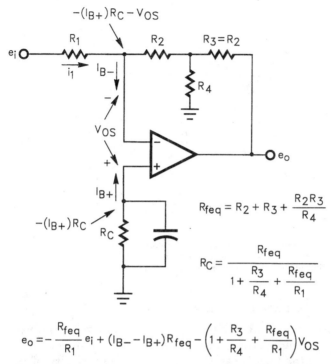

$$R_{feq} = R_2 + R_3 + \frac{R_2 R_3}{R_4}$$

$$R_C = \frac{R_{feq}}{1 + \dfrac{R_3}{R_4} + \dfrac{R_{feq}}{R_1}}$$

$$e_o = -\frac{R_{feq}}{R_1} e_i + (I_{B-} - I_{B+})R_{feq} - \left(1 + \frac{R_3}{R_4} + \frac{R_{feq}}{R_1}\right)V_{OS}$$

Figure 1.8 While the tee feedback reduces the offset effects of input bias currents, it potentially makes the op amp's input offset voltage a dominant effect, requiring a design compromise.

The resulting e_o reflects the tee's equivalent feedback resistance through $e_o = -(R_{feq}/R_1)e_i$, where $R_{feq} = R_2 + R_3 + R_2 R_3/R_4$. In practice, making $R_3 = R_2$ simplifies analysis by reducing the equivalent resistance to $R_{feq} = (2 + R_3/R_4)R_2$. Then, the tee effectively multiplies the resistance of R_2 by the factor $(2 + R_3/R_4)$.

Offset voltage analysis reveals the corresponding resistance reduction achieved for R_C using Fig. 1.8. Analogous to the basic inverting amplifier, I_{B-} flows through the feedback resistance, producing an output offset component of $I_{B-}R_{feq}$. Similarly, the flow of I_{B+} in R_C produces a counteracting component at the op amp's noninverting input equal to $-I_{B+}R_C$. To the latter voltage, the circuit appears as a noninverting amplifier, and at first, it might seem that the circuit would simply amplify that voltage by a gain of $1 + R_{feq}/R_1$. However, the grounded leg of the tee results in additional feedback current from this offset component and increases the corresponding gain to $1 + R_3/R_4 + R_{feq}/R_1$. Together, the two op amp input currents produce an output offset equal to $V_{OSO} = I_B\text{-}R_{feq} - (1 + R_3/R_4 + R_{feq}/R_1)I_B\text{+}R_C$. Then, assuming that $I_{B^-} = I_{B^+}$ and solving for R_C defines the optimum value for this resistance that makes $V_{OSO} \approx 0$. For this condition,

$$R_C = \frac{R_{feq}}{1 + R_3/R_4 + R_{feq}/R_1}$$

Then, ideal resistance matching produces $V_{OSO} = (I_{B^-} - I_{B^+})R_{feq}$.

A more detailed analysis next develops an optimum tee design by including the effects of the op amp's input offset voltage V_{OS}. The preceding discussion would suggest making the tee network's R_3/R_4 ratio large to reduce the required resistance values for R_2 and the associated R_C. That condition would absolutely avoid the higher tolerance errors of large resistance values. However, another offset component and noise complications impose a compromise in the R_3/R_4 selection. The tee also amplifies the op amp's input offset voltage, and this effect potentially overrides the offset reduction benefit of the tee network. For optimum offset reduction, a design equation guides a compromise selection of the R_3/R_4 ratio. In the figure, V_{OS}, like the $-I_{B^+}R_C$ component before, essentially drives a noninverting amplifier that supplies a gain of $1 + R_3/R_4 + R_{feq}/R_1$. Neglecting resistor mismatches, this makes the net output offset

$$V_{OSO} = (I_{B^-} - I_{B^+})R_{feq} - \left(1 + \frac{R_3}{R_4} + \frac{R_{feq}}{R_1}\right)V_{OS}$$

Minimization of this V_{OSO} requires avoiding an R_3/R_4 ratio that would make the V_{OS} component dominant over that produced by the I_B terms. With the tee network, the more accurate resistor matching achievable makes the I_B offset component largely a function of the matching of the I_{B^-} and I_{B^+} bias currents. Typically, these currents match to within 10% and produce a residual output offset component of $0.1(I_B R_{feq})$. Limiting the V_{OS} component of the equation to this same level produces a good compromise and defines a test for the choice between conventional and tee feedback connections. Under these conditions, the tee network benefits offset performance if

$$\frac{R_3}{R_4} \leq \left(\frac{0.1I_B}{V_{OS}} - \frac{1}{R_1}\right)R_{feq}$$

Then, choosing R_1 for the desired input resistance and choosing R_{feq} to set the $-R_{feq}/R_1$ gain defines a limit for the R_3/R_4 ratio in terms of the op amp's input bias current and offset voltage. Note that the application of this equation can produce an unexpected result that might confuse the tee design process. The minus sign of the equation can produce a negative result for the R_3/R_4 ratio limit expressed. Such results indicate that V_{OS} already dominates offset and the use of the tee can only degrade the overall result. For most high-gain applications, the equation delivers a positive result, for which the inverting amplifier with tee feedback delivers an output voltage of

$$e_o = -\frac{R_{feq}}{R_1} e_i + (I_{B^-} - I_{B^+})R_{feq} - \left(1 + \frac{R_3}{R_4} + \frac{R_{feq}}{R_1}\right)V_{OS}$$

1.2.3   Noninverting amplifiers with tee feedback

In general noninverting amplifiers do not require tee feedback to deliver both high input resistance and high gain. Unlike the inverting amplifier, the noninverting amplifier presents the op amp's high noninverting input impedance to the signal source, making the input impedance independent of the value of the conventional R_1 resistor. This permits lower R_1 values and corresponding lower values of R_2 for a given $1 + R_2/R_1$ gain. However, the occasional requirement for very high gain levels can encounter another resistance tolerance limit. There, keeping R_2 at a reasonable level would force R_1 to such a low level that parasitic series resistance limits the net R_1 tolerance and the resulting gain accuracy. Then, converting R_2 to a tee retains reasonable resistance values for this segment of the feedback network and permits the use of a larger resistance for R_1.

For offset and gain analysis of this case, Fig. 1.9 illustrates the noninverting amplifier with tee feedback replacing the conventional R_2 resistor. Superposition analysis first demonstrates that the offset response remains the same as developed previously for the inverting case. Comparing Figs. 1.8 and 1.9, superposition grounding of their e_i input terminals reduces them to the same circuit. Thus the noninverting amplifier with tee feedback produces the same output offset derived before of

$$V_{OSO} = (I_{B^-} - I_{B^+})R_{feq} - \left(1 + \frac{R_3}{R_4} + \frac{R_{feq}}{R_1}\right)V_{OS}$$

Once again, R_C meets the condition for minimum offset when

$$R_C = \frac{R_{feq}}{1 + R_3/R_4 + R_{feq}/R_1}$$

Also as before, converting R_2 to a tee reduces offset as long as

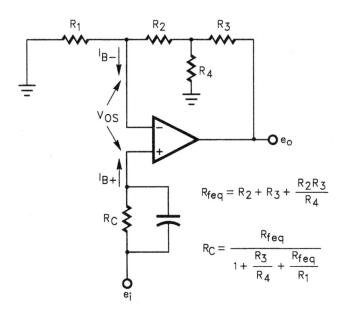

$$R_{feq} = R_2 + R_3 + \frac{R_2 R_3}{R_4}$$

$$R_C = \frac{R_{feq}}{1 + \dfrac{R_3}{R_4} + \dfrac{R_{feq}}{R_1}}$$

$$e_o = \left(1 + \frac{R_3}{R_4} + \frac{R_{feq}}{R_1}\right)(e_i - V_{OS}) + (I_{B-} - I_{B+})R_{feq}$$

Figure 1.9   While required less frequently, tee feedback with the noninverting configuration permits the combination of high gain and accurate R_C compensation without forcing R_1 to impractical low resistance levels.

$$\frac{R_3}{R_4} \le \left(\frac{0.1 I_B}{V_{OS}} - \frac{1}{R_1}\right) R_{feq}$$

Indirectly, this circuit's signal gain also follows from the foregoing inverting circuit discussion. At first, it might seem that the circuit in Fig. 1.9 would amplify e_i by a gain of $1 + R_{feq}/R_1$, paralleling the gain supplied by the basic noninverting amplifier. However, the grounded leg of the tee results in additional feedback current from e_i and increases the corresponding gain to $1 + R_3/R_4 + R_{feq}/R_1$. This repeats the gain described before for the inverting amplifier with tee feedback in response to the $-I_{B+}R_C$ offset component. Together, the offset errors and the input signal produce an output for the tee-feedback noninverting circuit equal to

$$e_o = \left(1 + \frac{R_3}{R_4} + \frac{R_{feq}}{R_1}\right)(e_i - V_{OS}) + (I_{B-} - I_{B+})R_{feq}$$

1.3   Differential-Input Amplifiers

Differential-input configurations add common-mode rejection to the op amp's signal processing function. They do so by combining components of inverting and noninverting gain to remove any signal common to the circuit's two inputs through subtraction. In the simplest case, just adding two resistors to the inverting op amp configuration produces the basic difference amplifier considered here. Multiple op amp configurations that form instrumentation amplifiers offer higher-performance differential-input results, as described in a later chapter. For the difference amplifier, common-mode rejection depends on two limits, as imposed by the finite rejection capability of the op amp and by the circuit's resistor mismatches. Analyses of these two limits define nonzero common-mode gains that set the final common-mode rejection result. Variations on the basic difference amplifier extend its input common-mode range to high voltages or permit simple common-mode rejection trimming that does not disturb gain accuracy.

1.3.1   Difference amplifiers

Unlike the preceding inverting and noninverting connections, the difference amplifier configuration realizes the full differential-input capability of an op amp. Previously, the inverting connection accepted an input signal at one of the op amp's two inputs and the noninverting connection accepted a signal at the other. The difference amplifier accepts signals at both inputs to activate the op amp's inherent common-mode rejection capability.

Shown in Fig. 1.10, the difference amplifier combines components of inverting and noninverting gain. There, matching resistor networks couple two input signals to the op amp's inputs, and superposition analysis of the circuit defines the resulting output signal. First, superposition grounding of the e_2 input permits analysis of the e_1 signal effect. This grounding places the lower R_1 resistance in parallel with its R_2 counterpart. As a side benefit, this parallel connection reveals that the matched resistor networks used here automatically produce the offset reduction previously achieved by adding $R_C = R_1 \| R_2$ in series with the op amp's noninverting input. With the e_2 input grounded, input signal e_1 drives an inverting amplifier to produce an output signal component of $e_o = -(R_2/R_1)e_1$.

Next, superposition grounding of the e_1 input terminal permits analysis of the e_2 signal effect in two steps. Signal e_2 first encounters an attenuation that delivers a signal equal to $e_2 R_2/(R_1 + R_2)$ at the op amp's noninverting input. There, the attenuated signal drives a noninverting amplifier and receives a gain equal to $1 + R_2/R_1$ to produce the output component $e_o = (R_2/R_1)e_2$. Thus e_2 transfers to the output with the same

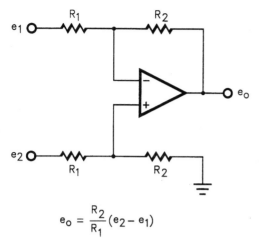

$$e_o = \frac{R_2}{R_1}(e_2 - e_1)$$

Figure 1.10 Adding a second resistor network to the inverting configuration forms the difference amplifier for rejection of common-mode input signals.

gain magnitude as e_1 but with opposite gain polarity. Together, the e_1 and e_2 input signals produce the output difference signal

$$e_o = \frac{R_2}{R_1}(e_2 - e_1)$$

As expressed, the difference amplifier both supplies voltage gain and subtracts one input signal from the other. This subtraction activates one of the more powerful capabilities of the op amp, common-mode rejection. Any signal common to e_1 and e_2 drops out in the subtraction process. The differential inputs of an op amp inherently produce this differencing action and the matched resistor networks of the difference amplifier configuration provide practical access to this capability.

The preceding equation defines the ideal voltage gain of the difference amplifier as $G_{da} = R_2/R_1$, which equals the gain magnitude produced by the inverting op amp configuration. Unfortunately, the factors underlying this gain magnitude also transfer the inverting configuration's error gain and gain–bandwidth disadvantages to the difference amplifier. As described in Sec. 1.1.2, the inverting configuration produces an error gain having the greater magnitude of $G_e = 1/\beta = 1 + R_2/R_1$. For the difference amplifier the upper resistor network continues to control the circuit's feedback factor $\beta = R_1/(R_1 + R_2)$, and the resulting error gain remains larger than the signal gain. Adding the lower resistor network to form the difference amplifier only introduces an input attenuator and does not alter the feedback factor from that of the basic inverting configuration. This distinction also makes the difference amplifier's gain–bandwidth product less than that of the op amp. As developed in Sec. 1.1, $BW = \beta f_c$ for op amp circuits where f_c is the unity-gain crossover of the op amp. For the difference amplifier, $BW_{da} = \beta f_c = f_c/(G_{da} + 1)$ and $GBW_{da} = G_{da}f_c/(G_{da} + 1)$. This GBW_{da} expression reflects that the difference amplifier does not realize the full gain–bandwidth product $GBW = f_c$ of the op amp. Compared with the noninverting configuration, the difference amplifier reduces GBW by a fraction equal to $G_{da}/(G_{da} + 1)$.

In addition, the difference amplifier presents input impedance conditions that restrict the amplifier's general use to lower source-impedance applications. First, the circuit presents unbalanced impedances at its two inputs, and this results in gain error. Superposition analysis first defines the impedance presented at the e_1 input by grounding the e_2 input. This reduces the circuit to an inverting amplifier having an input impedance equal to R_1. Next, superposition grounding of the e_1 input reduces the circuit to a noninverting amplifier with an input attenuator. This attenuator presents an input impedance equal to $R_1 + R_2$, which exceeds the R_1 impedance presented by the circuit's inverting input. Due to this impedance difference, the circuit's two inputs load their source connections differently and disturb the gain balance otherwise presented to e_1 and e_2. However, this input impedance imbalance does not disturb common-mode rejection. Setting $e_2 = e_1$ to illustrate the common-mode condition also establishes equal voltages across the upper and lower R_1 resistors so the circuit's two inputs then draw equal currents from their respective sources. With equal voltages and equal currents, the circuit's two inputs present equal impedances of $R_1 + R_2$ to common-mode signals. Thus no source loading imbalance disturbs common-mode rejection as long as equal source impedances drive the circuit's two inputs. In practice, matching the two source impedances to the degree required for the preservation of common-mode rejection requires that these impedances be low.

1.3.2   Difference amplifier CMRR

In the difference amplifier's subtraction process, practical limitations within the op amp and with the circuit's resistor matching retain a residual common-mode error. Analyses of the two error sources define the actual common-mode rejection of the difference amplifier by combining the effects of the two. For each analysis, derivations of the circuit's differential and common-mode gains define the limits imposed as guided by the definition of the common-mode rejection ratio (CMRR). Common-mode rejection equals the ratio of the circuit's differential and common-mode gain magnitudes, or $CMRR = |A_D/A_{CM}|$. Follow-

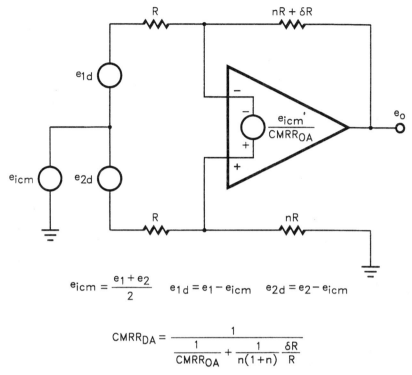

$$e_{icm} = \frac{e_1 + e_2}{2} \quad e_{1d} = e_1 - e_{icm} \quad e_{2d} = e_2 - e_{icm}$$

$$CMRR_{DA} = \cfrac{1}{\cfrac{1}{CMRR_{OA}} + \cfrac{1}{n(1+n)}\cfrac{\delta R}{R}}$$

Figure 1.11 Difference amplifier imposes limits to common-mode rejection through the op amp's inherent common-mode error and through the resistor mismatches consolidated here in δR.

ing the individual analyses of amplifier and resistor effects, a separate analysis combines the two to define the net CMRR of the difference amplifier.

Figure 1.11 models the difference amplifier for these analyses. There, the op amp presents its common-mode error in the differential input error signal $c'_{icm}/CMRR_{OA}$, as introduced earlier with Fig 1.1. Also, the model here consolidates the circuit's resistor matching error in the δR component shown. In practice, tolerance errors in each of the circuit's four resistors contribute to CMRR degradation, but combining their net effect in the δR component simplifies the analysis and delivers an equivalent result. This δR component includes any source resistance additions to the R_1 input resistors introduced by the application circuit. Generalizing the circuit's resistor values to R and nR, as shown, further simplifies the analysis. Also for simplicity, the model separates the common-mode component of the previous e_1 and e_2 input signals into an e_{icm} signal. Superposition analysis with the remaining e_{1d} and e_{2d} differential components first defines the circuit's differential gain A_D. For this analysis, superposition zeroing of e_{icm} returns the circuit to that of Fig. 1.10, where the circuit delivered a differential gain magnitude of $A_D = R_2/R_1$. For the model of Fig. 1.11, $R_1 = R$ and $R_2 \approx nR$, making $A_D = n$.

Given this differential gain, defining the common-mode rejection limits imposed by the op amp and the resistor matching only requires finding the associated common-mode gains. Analyses of these gains use superposition to zero of the e_{1d} and e_{2d} signals of the figure and isolate the output effects of e_{icm}. Consider first the limit imposed by the finite common-mode rejection of the op amp, CMRR_{OA}. For this analysis, neglecting the δR component focuses solely on the effect of the op amp without materially affecting the end result. Then, the e_{icm} input signal develops a signal equal to $e'_{icm} = e_{icm}n/(1 + n)$ at the op amp's noninverting input, and feedback essentially replicates this signal at the op amp's inverting input. This makes the op amp's common-mode input voltage the attenuated signal e'_{icm}, and the common-mode input error signal of the op amp becomes $e'_{icm}/\text{CMRR}_{\text{OA}} = e_{icm}n/(1 + n)\text{CMRR}_{\text{OA}}$. This error signal drives a noninverting amplifier that supplies a gain of $1 + n$ to produce an output signal equal to $e_o = e_{icm}n/\text{CMRR}_{\text{OA}}$ for a common-mode gain of $A_{\text{CM}} = e_o/e_{icm} = n/\text{CMRR}_{\text{OA}}$. Given this gain, the previous $\text{CMRR} = |A_D/A_{\text{CM}}|$ and $A_D = n$ reduce the CMRR limit imposed upon the difference amplifier by the op amp to CMRR_{OA}. Thus the op amp imposes a CMRR limit to the difference amplifier equal to the amplifier's inherent common-mode rejection ratio.

Next, consider the common-mode rejection limit imposed by resistor mismatch, as summarized in the δR component shown. That component actually represents the difference between the circuit's two $R{:}nR$ matching errors. As long as the actual $R{:}nR$ ratio of the upper resistor network matches that of the lower network, the circuit remains balanced and no common-mode error results from resistance mismatch. Only differences between these two matching ratios produce such an error, as summarized by the δR component. Then, $\delta R/nR$ represents the net fractional mismatch and $100\delta R/nR$ represents the net percentage mismatch. This mismatch disturbs the difference amplifier's subtraction process described before and introduces a common-mode error at the circuit's output. To separate the analysis of this effect, superposition zeroing of e_{1d}, e_{2d}, and $e'_{icm}/\text{CMRR}_{\text{OA}}$ in Fig. 1.11 leaves e_{icm} as the only signal at the circuit's input. Analysis of this condition reveals that nonzero values of δR result in an output error signal equal to $e_o = -e_{icm}\delta R/(1 + n)R$ for a common-mode gain of $A_{\text{CM}} = -\delta R/(1 + n)R$. Given this gain, the previous $\text{CMRR} = |A_D/A_{\text{CM}}|$ and $A_D = n$ reduce the CMRR limit imposed upon the difference amplifier by resistor mismatch to

$$\text{CMRR}_R = \frac{(1 + n)nR}{\delta R} = \frac{(1 + A_D)nR}{\delta R}$$

Note the sensitivity of this CMRR limit to the differential gain A_D. A given fractional mismatch of $\delta R/nR$ produces a CMRR_R limit that increases in proportion to $1 + A_D$. This reflects that increasing A_D improves CMRR_R by increasing the circuit's differential gain without changing its common-mode gain.

Together, the CMRR limits imposed by the op amp and by the resistor matching define the final CMRR of the difference amplifier. Directly combining the output error signals described for the two limits produces

$$e_o = \left(\frac{n}{\text{CMRR}_{OA}} - \frac{\delta R}{(1 + n)R} \right) e_{icm}$$

In this expression, the minus sign suggests that the error signals introduced by the op amp and the resistors tend to cancel. In some cases they do produce counteracting effects, but the uncertainty of the error polarity and the phase shift associated with CMRR_{OA} makes this unpredictable and requires replacing the minus sign with a plus for conservative analysis. The actual error produced by finite CMRR_{OA} can be of either polarity due to the differential nature of the op amp's inputs, and in 50% of practical cases these two error terms add rather than subtract. Further, the error signal resulting from CMRR_{OA} displays a 90° phase shift over most of the op amp's useful frequency range due to CMRR_{OA} response roll off.

To accommodate these polarity and phase ambiguities, the fundamental definition of $\text{CMRR} = |A_D/A_{CM}|$ incorporates an absolute value function. To accommodate this condition here, replacing the earlier minus sign with a plus and solving for CMRR yields a conservative expression for the difference amplifier,

$$\text{CMRR}_{DA} = \frac{1}{1/\text{CMRR}_{OA} + \delta R/(1 + A_D)nR} = \text{CMRR}_{OA} \parallel \text{CMRR}_R$$

When applying this equation, note that CMRR denotes the linear form of the expression and not the logarithmic form commonly defined as CMR, where $\text{CMR} = 20 \log(\text{CMRR})$.

1.3.3 High-voltage difference amplifiers

A voltage swing limitation of the op amp restricts the common-mode rejection feature of the basic difference amplifier to lower-voltage applications. As described, the basic configuration develops the signal

$e_2 R_2/(R_1 + R_2)$ at the op amp's noninverting input and feedback replicates this signal at the op amp's inverting input. For linear operation, this common-mode input signal must remain within the op amp's input voltage range, and this restricts the voltage ranges of both e_2 and the associated common-mode input signal $e_{icm} = (e_1 + e_2)/2$. For most op amps, this restriction precludes the monitoring of signals in the presence of high common-mode voltages, such as occur in current monitoring of an ac power line.

However, a simple modification to the difference amplifier extends its common-mode range to hundreds of volts as long as the accompanying compromises in accuracy and bandwidth remain acceptable. The key to this modification lies in the distinction between the actual common-mode input signal e_{icm} and that reaching the op amp's inputs, e'_{icm}. The actual signal $e_{icm} = (e_1 + e_2)/2$ inherently contains any high-voltage signal common to e_1 and e_2. However, only an attenuated portion of this signal reaches the op amp in $e'_{icm} = e_2 R_2/(R_1 + R_2)$. Then making $R_2 \ll R_1$ greatly attenuates a high-voltage signal, restraining e'_{icm} to the input voltage range of the op amp. This measure alone would restrict the difference amplifier's R_2/R_1 gain to levels well below unity except for the inclusion of a gain-restoring resistor, as shown in Fig. 1.12. There, first replacing the lower R_2 resistor of the basic difference amplifier with R_2/m reduces the voltage reaching the op amp's inputs to

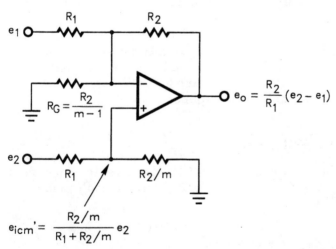

$$e'_{icm} = \frac{R_2/m}{R_1 + R_2/m} e_2$$

Figure 1.12 Increasing the attenuation of the e_2 input signal with the R_2/m resistor extends the circuit's common-mode input range, and compensating for the signal attenuation with the R_G resistor retains the circuit's differential gain.

$$e'_{icm} = \frac{R_2/m}{R_1 + R_2/m}\, e_2$$

Then, adding the R_G gain-restoring resistor compensates for the increased attenuation of the desired e_2 signal component. Superposition analysis of the circuit response defines the required R_G value and the relative reduction of e'_{icm}. As before, superposition grounding of the e_2 input reduces the circuit to an inverting amplifier that supplies a gain of $-R_2/R_1$ to e_1. This gain remains unaffected by the presence of the added R_G resistor. However, to retain the difference amplifier action, that resistor must ensure the same gain magnitude for e_2. Superposition analysis translates this requirement into an R_G resistor value by grounding the e_1 input. This reduces the circuit to a noninverting amplifier that supplies a gain of $1 + R_2/(R_1 \| R_G)$ to the attenuated e'_{icm} signal at the op amp's noninverting input. As expressed, the presence of R_G increases the gain supplied to that attenuated signal to permit a greater reduction of the common-mode signal coupled to the op amp's inputs. Together, this noninverting gain and the attenuation define the net gain magnitude supplied to e_2, and equating this magnitude to the previous R_2/R_1 defines the required R_G as

$$R_G = \frac{R_2}{m-1}$$

Establishing this value for R_G retains the $e_o = (R_2/R_1)(e_2 - e_1)$ response of the basic difference amplifier.

As before, this response reflects a differential gain magnitude of $A_D = R_2/R_1$, and this gain magnitude competes with the factor m in the reduction of the common-mode signal reaching the op amp. Dividing the original e'_{icm} of the basic difference amplifier by the modified e'_{icm} of the high-voltage alternative produces a signal reduction factor of

$$\frac{m + A_D}{1 + A_D}$$

As expressed, making $m > 1$ produces a signal reduction factor greater than unity. However, the relative reduction remains vulnerable to higher values of A_D. A combination of $A_D \gg m$ and $A_D \gg 1$ reduces this factor to essentially unity for zero reduction in the op amp's common-mode input signal. Practical implementations of the circuit do permit $A_D \gg 1$ as long as $A_D < m$, for which the reduction factor simplifies to $1 + m/A_D$. Then as this equation expresses, an increase in A_D requires a corresponding increase in m to retain a given reduction factor.

In practice, circuit accuracy and bandwidth requirements limit the practical magnitudes of both m and A_D through the effects of these parameters on the circuit's feedback factor. For single-feedback circuits like this, that factor equals the voltage divider ratio of the feedback network as seen between the op amp's output and its inverting input.[4] Superposition analysis of this factor grounds the e_1 input, placing R_1 in parallel with R_G, and the presence of R_G reduces the feedback factor from the original $R_1/(R_1 + R_2)$ to

$$\beta = \frac{R_1}{mR_1 + R_2} = \frac{1}{m + A_D}$$

Thus increasing either m or A_D reduces the circuit's feedback factor β.

As developed in Sec. 1.1.1, an op amp circuit amplifies its input-referred error sources by a gain of $G_e = 1/\beta$, and in this case $G_e = 1/\beta = m + A_D$. That gain amplifies the output errors resulting from the op amp's input offset voltage, input noise voltage, and input common-mode error signal. Indirectly, that gain also affects the circuit's bandwidth through the relationship BW $= \beta f_c = f_c/G_e$, as also developed in Sec. 1.1.1. In this case, that bandwidth relationship produces BW $= f_c/(m + A_D)$. Thus increasing m or A_D for either improved common-mode range or increased circuit gain also increases circuit errors and reduces signal bandwidth.

1.3.4   High-voltage difference amplifier CMRR

In spite of the increased error gain, the high-voltage difference amplifier does not reduce the common-mode rejection limit imposed by the op amp. However, this circuit does reduce the limit imposed by resistor mismatch. An analysis paralleling that of the basic difference amplifier defines the net result by combining the effects of these two common-mode error sources. For each source, derivations of the circuit's differential and common-mode gains define the limits imposed by each error source through the definition of the common-mode rejection ratio. Common-mode rejection equals the ratio of the circuit's differential and common-mode gain magnitudes, or CMRR $= |A_D/A_{CM}|$. Following the individual analyses, a separate analysis combines the two error effects to define the net CMRR of the high-voltage difference amplifier.

Figure 1.13 models this modified amplifier for CMRR analysis. As with the basic difference amplifier, the model represents the op amp's common-mode error signal with the differential-input source $e'_{icm}/$ CMRR_{OA} and consolidates resistor matching errors in a δR component. In practice, tolerance errors in each of the circuit's resistors contribute to CMRR degradation, but combining their net effect in the δR component simplifies the analysis and delivers an equivalent result. To further simplify analysis, the model also generalizes the circuit's basic R_1

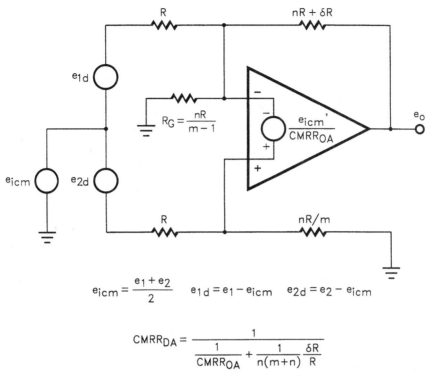

$$e_{icm} = \frac{e_1 + e_2}{2} \qquad e_{1d} = e_1 - e_{icm} \qquad e_{2d} = e_2 - e_{icm}$$

$$CMRR_{DA} = \cfrac{1}{\cfrac{1}{CMRR_{OA}} + \cfrac{1}{n(m+n)}\cfrac{\delta R}{R}}$$

Figure 1.13 High-voltage version of the difference amplifier retains the same common-mode rejection limit imposed by the op amp but reduces the limit imposed by δR due to the scaling factor m.

and R_2 resistor values to R and nR. Finally, the model separates the common-mode component of the e_1 and e_2 input signals into an e_{icm} signal. Superposition analysis with the remaining e_{1d} and e_{2d} differential-input components first defines the circuit's differential gain A_D. For this analysis, superposition zeroing of c_{icm} returns the circuit to that of Fig. 1.12, where the circuit delivered the differential gain magnitude $A_D = R_2/R_1$. In this case, $R_1 = R$ and $R_2 \approx nR$, making $A_D = n$.

Given this differential gain, defining the common-mode rejection limits imposed by the op amp and the resistor matching only requires finding the associated common-mode gains. Analyses of these gains use superposition to zero the e_{1d} and e_{2d} signals of the figure and isolate the output effect of e_{icm}. Consider first the limit imposed by the finite common-mode rejection of the op amp, $CMRR_{OA}$. For that case, neglecting the δR component focuses the analysis upon the effect of $CMRR_{OA}$ without materially affecting the end result. Then, the e_{icm} input signal develops a signal equal to $e_{icm}n/(m + n)$ at the op amp's noninverting input and feedback essentially replicates this signal at the op amp's inverting input. This makes the op amp's common-mode input voltage the attenuated signal $e'_{icm} = e_{icm}n/(m + n)$, and the common-mode input error signal of the op amp becomes $e'_{icm}/\text{CMRR}_{OA} = e_{icm}n/(m + n)\text{CMRR}_{OA}$. Neglecting δR, this error signal drives a noninverting amplifier, which supplies a gain of $m + n$ to produce an output signal equal to $e_o = e_{icm}n/\text{CMRR}_{OA}$ for a common-mode gain of $A_{CM} = e_o/e_{icm} = n/\text{CMRR}_{OA}$. Given this gain, the previous CMRR $= |A_D/A_{CM}|$ and $A_D = n$ simplify the common-mode rejection limit imposed upon the high-voltage difference amplifier by the op amp to CMRR_{OA}.

This result repeats that of the basic difference amplifier where the op amp imposes a common-mode rejection limit equal to that amplifier's inherent common-mode rejection ratio. Two effects make this result the same for the high-voltage case. First, the circuit's added input attenuation reduces e'_{icm} and the resulting $e'_{icm}/\text{CMRR}_{OA}$ error signal, and this tends to improve the circuit's common-mode rejection. Second, the addition of R_G increases the gain supplied to this reduced error signal, and this tends to degrade the rejection. In practice, selecting R_G to balance the circuit's inverting and noninverting gains also balances these two common-mode rejection effects for a net zero change in the op amp's effect.

Next consider the common-mode rejection limit imposed by resistor mismatch, as summarized in the δR component shown. That component actually represents the net matching errors of the circuit's multiple-resistor matching requirements. These requirements include the $nR{:}R$, $(nR/m){:}R$, and $[nR/(m - 1)]{:}R$ matching ratios expressed in the figure. The individual matching errors of these ratios produce common-mode rejection effects that may either add to or subtract from the net matching error, depending on mismatch polarities. While this produces a seemingly complex error combination, the three matching ratios contain a common factor nR that permits a simple representation of the net effect. Then $\delta R/nR$ represents the net fractional mismatch, and $100\delta R/nR$ represents the net percentage mismatch. For a given application, combining the individual percentage errors of the three matching ratios identified and equating the net result to $100\delta R/nR$ defines δR.

This mismatch disturbs the difference amplifier's subtraction process, producing a common-mode error at the circuit output. To separate the analysis of this effect, superposition zeroing of e_{1d}, e_{2d}, and $e'_{icm}/\text{CMRR}_{OA}$ in Fig. 1.13 leaves e_{icm} as the only signal at the circuit's input. Analysis of this condition reveals that nonzero values of δR result in an output error signal equal to $e_o = -e_{icm}\delta R/(m + n)R$ for a common-mode gain of $A_{CM} = -\delta R/(m + n)R$. Given this gain, the previous CMRR $= |A_D/A_{CM}|$ and $A_D = n$ reduce the common-mode rejection limit imposed upon the high-voltage difference amplifier by resistor mismatch to

$$\mathrm{CMRR}_R = \frac{(m + n)nR}{\delta R} = \frac{(m + A_D)nR}{\delta R}$$

This result differs from the basic difference amplifier case only in the m terms, which replace the "1" terms from before. Note the sensitivity of this CMRR limit to the factor m and the differential gain A_D. A given fractional mismatch of $\delta R / nR$ produces a CMRR_R limit that increases in proportion to $(m + A_D)$. Increasing A_D improves CMRR_R by increasing the circuit's differential gain without changing its common-mode gain. Conversely, increasing m decreases the circuit's common-mode gain without changing its differential gain.

Together, the common-mode rejection limits imposed by the op amp and by the resistor matching define the final CMRR of the difference amplifier. As described for the basic difference amplifier, the two limits combine in a manner that replicates the parallel combination of two resistances, and for the high-voltage difference amplifier,

$$\mathrm{CMRR}_{DA} = \frac{1}{1/\mathrm{CMRR}_{OA} + \delta R/(m + A_D)nR} = \mathrm{CMRR}_{OA} \, \| \, \mathrm{CMRR}_R$$

For application of this equation, note that CMRR represents the linear form of the expression and not the logarithmic form commonly defined as CMR, where CMR = 20 log(CMRR).

1.3.5 Adding CMRR trim

The two difference amplifiers of the preceding sections deliver common-mode rejection performance dictated by both the op amp's basic common-mode rejection and the resistor mismatch. At lower frequencies, the op amp's common-mode rejection typically remains very high, relegating the control of this characteristic to the resistor mismatch. There, resistor trim routinely increases low-frequency common-mode rejection beyond that of the initial assembly result. To confine this trim operation to just one resistor, adding a low-resistance potentiometer in series with that resistor and undersizing that resistor's value by one-half the potentiometer's resistance permits a bipolar trim. Then the potentiometer adjustment controls the net resistance match in either direction about an ideal null that optimizes common-mode rejection.

This adjustment serves difference amplifiers built from individual components, but the fixed resistances of preassembled difference amplifiers preclude this option. In such cases no undersizing of a given resistor's value permits a bipolar trim with just the addition of a potentiometer. Instead, these amplifiers would require an additional offsetting

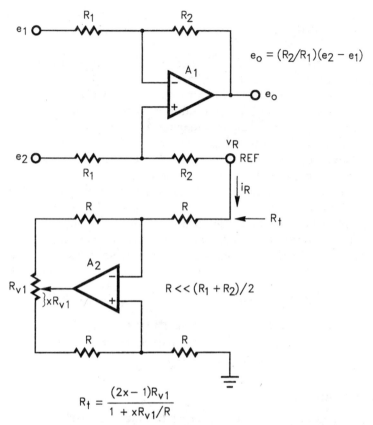

Figure 1.14 For the difference amplifier, adding a second op amp having both positive and negative feedback provides a CMRR trim resistance R_t that is variable from positive values through zero to negative values.

resistor, which disturbs gain accuracy. To remove this constraint, a simple op amp circuit converts the normal unipolar resistance control of a potentiometer to bipolar operation.[5] Shown in Fig. 1.14 with the basic difference amplifier, the CMRR trim circuit consists of A_2 and its associated feedback elements. These elements supply both negative and positive feedback around A_2 as controlled by potentiometer R_{v1}. Varying R_{v1} alters the relative amounts of the two feedback signals to control both the magnitude and the polarity of the impedance presented to the reference pin (REF) of the difference amplifier. Most applications of the difference amplifier simply connect this reference pin to ground to produce a ground-referenced output signal.

However, in this case the reference pin serves to balance the difference amplifier's resistances for trim of common-mode rejection. To examine the impedance control of the trim circuit, consider the voltage

v_R as developed by current i_R at the reference pin under different circuit conditions. First, consider the trim circuit without the positive feedback supplied by the lower left R resistor. Then, the trim circuit appears as a simple inverting amplifier with negative feedback and a virtual ground at A_2's inverting input. Because of this virtual ground, the circuit would then present an impedance equal to R at the reference pin and the current i_R would cause a rise in v_R. This voltage–current relationship describes a positive polarity for the R_t resistance presented by the trim circuit to the difference amplifier's reference pin.

Next, include the effect of the positive feedback of the lower left R resistor upon this impedance. When dominant, the positive feedback causes the output of A_2 to swing negative in response to the current i_R. Then, positive feedback couples a portion of this negative swing to the noninverting input of A_2, causing a matching negative swing at that amplifier's inverting input. Thus the A_2 inverting input no longer presents a virtual ground and the negative swing of the example counteracts the voltage rise previously described for v_R. In fact, with sufficient positive feedback this counteracting effect produces a negative swing for v_R in response to i_R. Such a voltage–current relationship describes a negative polarity for the R_t resistance presented to the reference pin.

Varying R_{v1} adjusts the relative amounts of negative and positive feedback controlling the trim circuit to vary the R_t trim resistance presented from positive values, through zero to negative values for bipolar control of CMRR. In the trim adjustment, making the R_{v1} setting fraction $x > 0.5$ produces more negative than positive feedback for A_2 and makes $R_t > 0$. Similarly, $x < 0.5$ makes $R_t < 0$, and $x = 0.5$ produces $R_t = 0$. Analysis defines the trim resistance control through the expression

$$R_t = \frac{(2x - 1)R_{v1}}{1 + xR_{v1}/R}$$

Adding the CMRR trim circuit to the difference amplifier leaves the basic circuit performance unchanged but introduces additional errors. The circuit's basic response remains $e_o = (R_2/R_1)(e_2 - e_1)$, and the signal bandwidth remains $\mathrm{BW}_{da} = \beta f_c = f_c/(G_{da} - 1)$. However, the trim circuit addition introduces offset, noise, and, potentially, high-frequency CMRR disturbance. The input offset voltage and input noise voltage of A_2 transfer directly to the difference amplifier output with unity gain. Fortunately, these errors add to the intended signal following the gain supplied by the difference amplifier and may remain insignificant. More significantly, the lower-frequency CMRR compensation provided by the trim amplifier may turn into a CMRR disturbance at higher frequencies. The compensating impedance presented by this amplifier inherently varies as the A_2 circuit approaches its bandwidth limit. For this reason, the bandwidth of the trim circuit must be made large compared to that of the difference amplifier. Achieving this result begins with the selection of a higher-frequency op amp for A_2.

However, the final result also requires attention to the feedback factor β_2 of the A_2 circuit connection. As described in Sec. 1.1.1, the trim amplifier delivers a bandwidth equal to $\beta_2 f_{c2}$, where f_{c2} is the unity-gain crossover frequency of A_2. Both f_{c2} and β_2 must be controlled to deliver the required trim circuit bandwidth. Fortunately, a simple resistor selection process assures that the trim circuit's negative feedback factor remains greater than the positive feedback factor to maximize β_2 and optimize this bandwidth. A simplified analysis defines the selection requirement by neglecting the fine-tuning effects of R_{v1}. For this analysis, setting $R_{v1} = 0$ simplifies the voltage dividers that define the circuit's positive and negative feedback factors.

As developed in another publication,[6] the feedback factor of a given feedback network equals the voltage divider ratio of that network as seen from the amplifier output. As also developed there, a circuit having both positive and negative feedback produces a net feedback factor of $\beta = \beta_- - \beta_+$. The trim circuit's own resistors define the positive feedback factor and, with $R_{v1} = 0$ for this analysis, $\beta_+ = R/(R + R) = 1/2$. This high level of positive feedback could produce oscillation. However, the negative feedback path of the trim circuit includes the lower R_2 and R_1 resistors of the difference amplifier in its feedback divider. This makes the negative feedback factor $\beta_- = (R + R_1 + R_2)/(2R + R_1 + R_2)$. Selecting $R \ll (R_1 + R_2)/2$ increases this factor to approximately $\beta_- = 1$ for a net feedback factor of $\beta = \beta_- - \beta_+ = 1/2$. This optimizes the trim circuit bandwidth $BW_2 = f_{c2}/2$. While this optimization generally requires a low resistance value for R, in practice only very small voltage swings develop across the R resistors, and they produce only small current drains.

References

1. J. Graeme, *Optimizing Op Amp Performance*, McGraw-Hill, New York, 1997.
2. Graeme, op cit., 1997.
3. Graeme, op cit., 1997.
4. Graeme, op cit., 1997.
5. J. Graeme, "Active Potentiometer Tunes Common-Mode Rejection," *Electronics*, p. 119, June 30, 1982.
6. Graeme, op cit., 1997.

2

Gain Options

The preceding chapter describes the fundamental voltage gain alternatives realized with op amps in noninverting, inverting, and difference configurations. For each of these, variations in this chapter produce gain options in the form of signal summation, variable gain, or switched gain. Signal summation provides weighted gains to a number of input signals, producing a composite output signal. Summation normally connotes just the inverting configuration, and there a well-known gain equation quickly defines the circuit's required resistor values. However, this configuration imposes a multielement voltage divider in the determination of the circuit's feedback factor for performance analysis. The lesser used, noninverting version of the summing amplifier imposes a similar multielement analysis at the outset in the determination of the gain-setting resistors. However, this configuration reduces the feedback factor calculation to that of a simple voltage divider ratio. In the end, the inverting and noninverting configurations present the same analytical challenges and analyses here reduce the tasks to simple design equations. The difference amplifier configuration imposes multielement analyses for the determination of both the resistor values and the feedback factor, but again, analyses presented here reduce these tasks to simple design equations.

Variable and switched gains result from potentiometers, analog multipliers, or switches connected in the op amp's feedback or input paths. In the simplest case, just replacing one of the circuit's traditional feedback resistors with a potentiometer produces continuously variable gain for the inverting and noninverting configurations. Different implementations of this approach develop either nonlinear or linear gain control characteristics. For the difference amplifier configuration, the preservation of common-mode rejection requires a carefully balanced connection of the potentiometer between the circuit's inverting and noninverting resistor networks. As a first alternative to the potentiometer, connecting an analog multiplier in the feedback path produces an effective variable resistance for electronic control of gain. As a second alternative, connecting feedback switches that access multiple feedback configurations also provides electronic control, but with discrete steps rather than continuously variable action.

2.1 Summing Amplifiers

Signal summation can be performed with each of the inverting, non-inverting, and difference amplifier configurations presented in the previous chapter. The three configurations produce similar results, with their primary differences residing in the resulting gain polarities and the design equations required. Summation with op amp circuits normally implies the inverting configuration, where simple math divisions define the weighted gains supplied to various summing inputs. However, this configuration presents a complex feedback factor analysis determined by a multielement voltage divider. As described in Sec. 1.1, that factor must be determined to define the circuit's error and bandwidth performance. Summation with the noninverting configuration extends the function to positive gain polarities and reduces the feedback factor computation to a two-element voltage divider. However, this configuration introduces a similar analytical complication through a multielement voltage divider in the gain selections. Analyses remove these complications for both configurations through a summation of the individual gains, and the resulting design equations make the noninverting configuration an equal contender for summing applications. Later analysis even shows the noninverting configuration to be somewhat superior in bandwidth and error gain.

Signal summation with the difference amplifier combines addition and subtraction in one amplifier stage. When configured appropriately, this amplifier also permits the summation or subtraction of differential signals. However, the difference amplifier introduces multielement voltage dividers that determine both the individual gains and the feedback factor. Once again, analyses reduce the complexities of the associated calculation tasks to simple algebra. The most practical implementation of this circuit addresses the summation of differential signals and retains the difference amplifier's unique common-mode rejection capability.

2.1.1 Inverting summing amplifiers

Signal summation with the inverting configuration presents the simplest mathematical determination of signal gains but requires a more complex analysis to determine the circuit's feedback factor. Figure 2.1

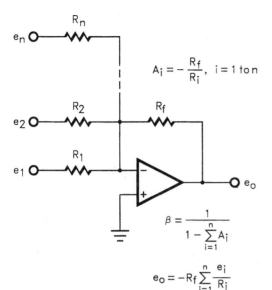

$$A_i = -\frac{R_f}{R_i}, \quad i = 1 \text{ to } n$$

$$\beta = \frac{1}{1 - \sum\limits_{i=1}^{n} A_i}$$

$$e_o = -R_f \sum_{i=1}^{n} \frac{e_i}{R_i}$$

Figure 2.1 Signal summation with the inverting configuration permits simple, independent gain calculations but requires a more complex calculation for the circuit's feedback factor.

illustrates this circuit for input signals e_1 to e_n with corresponding gains of A_1 to A_n. The virtual ground of this circuit's noninverting input reduces a given signal's gain calculation to the familiar

$$A_i = -\frac{R_f}{R_i}, \qquad i = 1, ..., n$$

and the amplifier produces a composite output signal of

$$e_o = -R_f \left(\frac{e_1}{R_1} + \frac{e_2}{R_2} + \cdots \frac{e_n}{R_n} \right)$$

This appealingly simple gain relationship leads most signal sum-mation applications to the inverting amplifier. However, this configu-ration presents a more complex condition for performance analysis due to the numerous resistors that influence the circuit's feedback factor. That factor determines the bandwidth and error signal gain for any op amp circuit as described in Sec. 1.1. Following a common rule-of-thumb guideline, the preceding gain expression might lead one to expect a bandwidth of around f_c/A_i for a given signal channel, where f_c is the unity-gain crossover frequency of the op amp. For a single signal input, the constant gain–bandwidth product of most op amps makes this rule-of-thumb relationship approximately true. However, the circuit's feedback factor, and not closed-loop gain, defines the as-sociated bandwidth.[1] From Sec. 1.1, an op amp circuit's bandwidth becomes BW = βf_c, and all signal channels of the summing amplifier experience this same bandwidth limit. Also from that chapter, the gain supplied to the amplifier's input error signals is $G_e \approx 1/\beta$, where β is the circuit's feedback factor. This gain amplifies the input errors associated with offset voltage, noise, and power-supply rejection. As a result of these two feedback consequences, signal summation typically serves only a limited number of summation inputs. When extended to a greater number, the summing amplifier can restrict bandwidth severely and increase error gain to high levels.

The summing inverting amplifier complicates the determination of β for these performance analyses but mathematical reduction produces a simple design equation for this purpose. As described in Sec. 1.1, the feedback factor β equals the voltage divider ratio of the feedback network as seen from output to input. The summing amplifier here complicates the calculation of this ratio with an arbitrary number of input summing resistors R_1 through R_n. This places an arbitrary number of resistors, in parallel, at the input side of the voltage divider. However, analysis reveals a simplified result expressed in terms of the summation of the individual channel gains and

$$\beta = \frac{1}{1 - \sum_{i=1}^{n} A_i}$$

Then, simply adding the gains of the various summing inputs and substituting the result in this equation defines β for bandwidth and error gain analyses. Note that this summation of negative gains produces a negative result for a net positive denominator in this equation.

2.1.2 Noninverting summing amplifiers

Generally overlooked in signal summation, the noninverting configuration initially presents a more complex gain calculation, but then a simpler feedback factor calculation. The net math requirement remains the same as with the inverting summing amplifier, and the analyses that follow reduce the complex gain calculations to a simplified design equation. In addition, the noninverting configuration offers positive gain polarities and somewhat greater bandwidth, as described in the previous chapter. The bandwidth advantage actually makes the noninverting amplifier superior to the inverting alternative in most summing applications. In return for these benefits, the noninverting summing amplifier requires just one additional circuit element, resistor R_G, as described next.

Shown in Fig. 2.2, the noninverting summing amplifier accepts input signals e_1 through e_m with corresponding gains of A_1 through

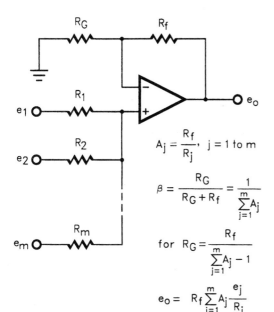

$$A_j = \frac{R_f}{R_j}, \quad j = 1 \text{ to } m$$

$$\beta = \frac{R_G}{R_G + R_f} = \frac{1}{\sum\limits_{j=1}^{m} A_j}$$

$$\text{for } R_G = \frac{R_f}{\sum\limits_{j=1}^{m} A_j - 1}$$

$$e_o = R_f \sum\limits_{j=1}^{m} A_j \frac{e_j}{R_j}$$

Figure 2.2 Addition of gain compensation resistor R_G permits noninverting summation with the simple gain calculations normally associated with the inverting summer.

A_m. There the input resistors R_1 through R_m combine to form a summing voltage divider, and each summing input added to the circuit complicates the individual divider ratios. This voltage divider inherently attenuates the signals at the amplifier input, requiring amplification to restore the output signal level. Adding the R_f/R_G feedback network produces this gain and provides an extra degree of design freedom to reduce the complexity of the gain selection. For a given input signal, the circuit produces a gain of

$$A_j = \left(1 + \frac{R_f}{R_G}\right) \frac{R_p}{R_p \mid R_j}, \quad j = 1, \ldots, m$$

where R_p is the parallel combination of the resistances R_1 through R_m but excluding the selected R_j. The multiple calculations of R_p required by this basic equation make it a poor design tool.

However, the factor R_f/R_G remains arbitrary and permits reducing the equation to

$$A_j = \frac{R_f}{R_j}, \quad j = 1, \ldots, m$$

This setting reproduces the gain selection convenience of the inverting case, given the appropriate selection of the R_f/R_G ratio. Together, R_f

and R_G present a load to the amplifier output, and selecting R_f for a convenient value limits the associated load current. To define R_G next, equating the two preceding expressions for A_j and solving produces

$$R_G = \frac{R_f}{\displaystyle\sum_{j=1}^{m} A_j - 1}$$

Then, simply adding the gains of the various summing inputs and substituting the result in the equation defines the value for R_G that establishes the $A_j = R_f/R_j$ condition.

This R_G value selection also defines the circuit's feedback factor for bandwidth and error analysis. The voltage divider ratio of the feedback network defines this factor as $\beta = R_G/(R_G + R_f)$ and substitution of the R_G result produces

$$\beta = \frac{1}{\displaystyle\sum_{j=1}^{m} A_j}$$

This result reflects the fundamental $\beta = 1/A$ relationship characteristic of the noninverting configuration. Comparison of this feedback factor result with that of the previous inverting summing amplifier indicates a somewhat larger β, and thereby lower error gain $G_e = 1/\beta$, for the noninverting summing amplifier. Also, this noninverting alternative delivers somewhat greater bandwidth, as given by the classic $BW = \beta f_c$ expression.

2.1.3   Difference summing amplifiers

As with the basic difference amplifier, the summing alternative combines characteristics of the inverting and noninverting counterparts. This combination permits signal addition as well as subtraction with individually weighted gains but introduces multielement voltage dividers for both the gain selections and the feedback factor calculation.[2] Analyses again reduce these multielement effects to simple design equations. Still, the difference amplifier's combination produces an interaction between inverting and noninverting gains. To counteract this, further analysis defines a gain compensation resistor that will generally be required in one or the other of two circuit locations.

Shown in Fig. 2.3, the difference summing amplifier delivers inverting gains for signals e_1- through e_n- and noninverting gains for e_1+ through e_m+. Resistors R_1- through R_n- set the gain weighting for the inverting side, and R_1+ through R_m+ set the weighting for the noninverting side. For the inverting side, the gain results remain the same

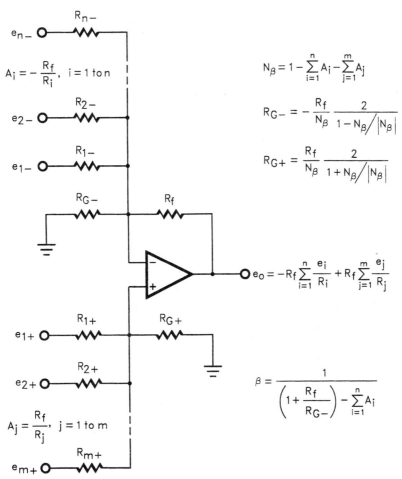

Figure 2.3 Adding either R_{G-} or R_{G+} extends simple gain calculations to the summing difference amplifier for the addition and subtraction of an arbitrary number of signals.

as described for the inverting summing amplifier. Superposition analysis of this gain sets e_{1+} through e_{m+} to zero, leaving an inverting summing amplifier with various grounded resistors at the op amp's noninverting input. Those resistors do not affect the inverting gains and, as before,

$$A_i = -\frac{R_f}{R_{i-}}, \qquad i = 1, ..., n$$

However, in the opposite superposition case, the resistors of the inverting side do affect the circuit's noninverting gains. In that case, superposition analysis sets e_1- through e_{n-} to zero, effectively grounding the corresponding resistors of the circuits' inverting inputs. This condition produces a noninverting amplifier with gains dependent on the R_i- resistors of the inverting side. This gain interaction generally requires the addition of either the R_G- or the R_G+ compensation resistor shown to restore independent control over the signal gain weightings of the inverting and noninverting inputs. That resistor also provides an extra degree of design freedom to permit a convenient gain-setting equation for noninverting inputs as expressed by

$$A_j = \frac{R_f}{R_{j+}}, \qquad j = 1, \ldots, m$$

Establishing this A_j relationship simply requires the appropriate choice of the R_G- or R_G+ resistor. The R_G resistor chosen must compensate for the independently determined gain effect that the R_i- resistors impose upon the noninverting e_{j+} signals. Those signals develop an attenuated sum at the op amp's noninverting input through the R_{j+} resistors. The circuit then amplifies this sum by a gain dependent on the R_i- resistors. Thus the two resistor sets both influence the gain supplied to a given noninverting input signal. While seemingly complex, analysis again reduces the gain setting and feedback factor calculations to simple algebra. This analysis first defines a fortuitous constant for a given circuit

$$N_\beta = \frac{1}{\beta_{is}} - \frac{1}{\beta_{ns}}$$

where β_{is} and β_{ns} are the feedback factors for the equivalent inverting and noninverting summing amplifiers that form the difference summing amplifier. Substitution of earlier β results for the inverting and noninverting summers in this equation gives

$$N_\beta = 1 - \sum_{i=1}^{n} A_i - \sum_{j=1}^{m} A_j$$

Then for a given circuit, simply adding the negative A_i gains, adding the positive A_j gains, and substituting the results in the preceding equation defines N_β. Note that gain polarities must be observed in this process and the $-\Sigma A_i$ term produces a positive number while the $-\Sigma A_j$ term produces a negative one.

Given N_β, the calculation of R_G- and R_G+ for a given circuit reduces to simple design equations. Establishing the $A_j = R_f/R_{j+}$ relationship requires that

$$R_{G^-} = -\frac{R_f}{N_\beta}\left(\frac{2}{1 - N_\beta/|N_\beta|}\right)$$

and

$$R_{G^+} = \frac{R_f}{N_\beta}\left(\frac{2}{1 + N_\beta/|N_\beta|}\right)$$

Then the presence of R_{G^-} compensates for too small a gain otherwise delivered to the e_{j^+} signals or the presence of R_{G^+} compensates for too large a gain. In most applications, one or the other of the two gain conditions prevails, requiring either R_{G^-} or R_{G^+}, but never both. Typically, the preceding equations deliver the desired result for the required resistor and an infinite result for the other. In some cases, balanced circuit conditions produce infinite results with both the R_{G^-} equation and the R_{G^+} equation, indicating no requirement for either resistor.

Selection of the R_G resistor defines the circuit's feedback factor for error and stability analyses. This factor follows from the inverting summing amplifier with a modification that accommodates the possible inclusion of an R_{G^-} resistor here. From before, the inverting summing amplifier develops a feedback factor of

$$\beta = \frac{1}{1 - \sum_{i=1}^{n} A_i}$$

The potential inclusion of the R_{G^-} resistor in the difference summing amplifier modifies this equation, producing

$$\beta = \frac{1}{(1 + R_f/R_{G^-}) - \sum_{i=1}^{n} A_i}$$

As discussed with the preceding design equations, the R_{G^-} resistance required may be infinite, leaving the feedback factor unchanged from that of the inverting summer.

2.1.4 Summing differential signals

The preceding difference amplifier adds or subtracts arbitrary numbers of input signals, and this well serves general applications. However, this general solution neglects a primary benefit of the difference amplifier, common-mode rejection. As discussed with the basic difference amplifier in Chap. 1, this amplifier rejects any signal com-

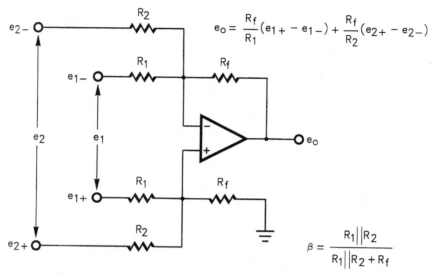

Figure 2.4 Restoring resistor balance to the difference summing amplifier of Fig. 2.3 reactivates common-mode rejection for the summation of differential signals.

mon to the two inputs supplied by a differential input signal. With balanced circuit conditions, the difference summing amplifier retains this rejection feature.

To illustrate differential signal summation and the circuit conditions required, consider the summation of two differential signals, as illustrated in Fig. 2.4. There, differential signals e_1 and e_2 drive the input connections of a balanced difference amplifier in a differential manner with ground-referenced signal components e_1+, e_1-, e_2+, and e_2-. As illustrated, these signal components form the differential signals $e_1 = e_1+ - e_1-$ and $e_2 = e_2+ - e_2-$. In the ideal case, each ground-referenced signal represents half the magnitude of the associated differential signal, making $e_1- = -e_1+$ and $e_2- = -e_2+$. However, in practice common-mode error signals disturb the magnitude equalities, potentially coupling common-mode error to the circuit output. Circuit balance prevents this coupling by establishing the common-mode rejection described in Sec. 1.3.2. For a given input signal, equal summing resistances, such as the R_1 and R_2 resistor pairs shown here, balance the circuit input. Further, making the previous R_G+ equal to R_f balances the voltage divider effects of the circuit's two resistor networks.

Note that this balance does not require making $R_1 = R_2$, and this design freedom retains the gain-weighting option of the previous summing amplifiers. As a result, this circuit configuration produces an output signal proportional to the sum of two differential signals as expressed by

$$e_o = \frac{R_f}{R_1}(e_1{}^+ - e_1{}^-) + \frac{R_f}{R_2}(e_2{}^+ - e_2{}^-)$$

It also produces a simplified feedback factor controlled by the inverting resistor network as expressed by

$$\beta = \frac{R_1 \| R_2}{R_1 \| R_2 + R_f}$$

This signal summation can obviously be extended to any number of differential input signals by simply adding summing resistor pairs to the circuit's inputs. For example, adding a third input signal e_3 to a pair of R_3 input resistors would add the term $(R_f/R_3)(e_3{}^+ - e_3{}^-)$ to the e_o expression and would replace the $R_1 \| R_2$ terms of the β expression with $R_1 \| R_2 \| R_3$.

2.2 Variable-Gain Amplifiers

Combining potentiometer or multiplier feedback with the basic inverting, noninverting, and difference op amp configurations produces continuously variable gain control. The next section develops switched-gain control configurations that produce discrete rather than continuous gain variation. The continuously variable alternatives described here produce both nonlinear and linear control characteristics with differing degrees of circuit complexity. In the simplest case, just an op amp and a potentiometer produce variable gain for both the inverting and the noninverting configurations. This combination produces a nonlinear control response, but simply adding a fixed resistor to the feedback linearizes the control for both configurations. Variable gain for the difference amplifier presents an additional challenge in maintaining the precise resistance matching that delivers common-mode rejection. There, connecting the control potentiometer in a differential tee network preserves the matching requirements and varies gain by shunting differential signal currents between the inverting and the noninverting sides of the circuit. An extension of the difference amplifier provides bipolar gain control that accesses both gain polarities under the control of one potentiometer. Finally, replacing the traditional feedback potentiometer with a fixed resistor and an analog multiplier makes the circuit's effective feedback resistance a function of a control voltage for electronic control of continuously variable gain.

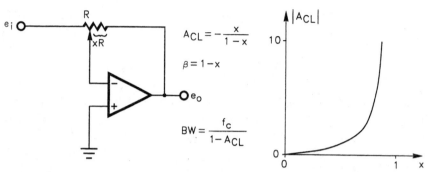

Figure 2.5 Just an op amp and a potentiometer produce variable inverting gain but with a nonlinear response having greater control sensitivity at low gain levels.

2.2.1 Variable inverting and noninverting gain

In the simplest configuration, just an op amp and a potentiometer produce variable gain, as illustrated by the inverting amplifier of Fig. 2.5. There, the wiper of the potentiometer serves as the feedback point at the op amp's inverting input and varying the potentiometer fraction x controls the circuit's closed-loop gain. Simple analysis defines the gain control response as

$$A_{CL} = -\frac{x}{1-x}$$

This response produces the nonlinear $|A_{CL}|$ versus x curve shown with a greater control sensitivity at lower gain levels, as often desired in audio applications.

The practical range of this variable gain depends on an application's bandwidth and accuracy requirements that ultimately define the circuit's minimum allowable feedback factor. Feedback through the potentiometer establishes a feedback factor of $\beta = 1 - x$ that potentially varies from one to zero as x ranges from zero to one. From Sec. 1.1, this factor controls bandwidth as described by $BW = \beta f_c$, where f_c is the unity-gain crossover frequency of the op amp. Thus this variable-gain circuit produces a variable bandwidth of

$$BW = (1-x)f_c = \frac{f_c}{1-A_{CL}}$$

Then the minimum bandwidth requirement for a given application defines a minimum β, a maximum x, and the circuit's maximum gain magnitude of

$$|A_{CL}|_{max} = \frac{f_c}{BW_{min}} - 1$$

Also from Sec. 1.1, the feedback factor defines the circuit's error gain as $G_e = 1/\beta$ over the circuit's usable bandwidth range. This gain amplifies the op amp's input-referred error sources such as offset, noise, CMRR error, and PSRR error. As a result, gain G_e places a second limit upon the maximum practical gain of this circuit. To limit the error gain to a given $G_{e\,max}$, the circuit's closed-loop gain must be constrained to

$$|A_{CL}|_{max} = |G_e|_{max} - 1$$

The noninverting counterpart of this circuit produces similar but somewhat superior results. As shown in Fig. 2.6, this circuit offsets the A_{CL} response upward due to the characteristic "1+" term of a noninverting circuit's fundamental $1 + R_2/R_1$ gain equation. Otherwise the nonlinear shape of the response remains the same as described by

$$A_{CL} = \frac{1}{1 - x} = \frac{1}{\beta}$$

Note that $\beta = 1 - x$ also remains the same, as controlled by the feedback voltage divider ratio of the potentiometer. However, the numerator of the A_{CL} expression changes, again due to the characteristic "1+" term, and this increases the maximum practical gain limits to

$$A_{CL\,max} = \frac{f_c}{BW_{min}}$$

and

$$A_{CL\,max} = G_{e\,max}$$

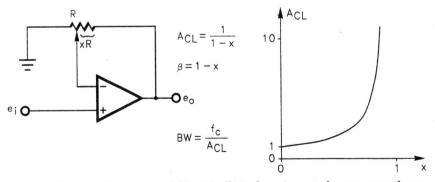

$$A_{CL} = \frac{1}{1 - x}$$

$$\beta = 1 - x$$

$$BW = \frac{f_c}{A_{CL}}$$

Figure 2.6 Noninverting version of Fig. 2.5 offsets the gain control curve upward, producing a minimum gain of unity, but otherwise the shape of the curve remains unchanged.

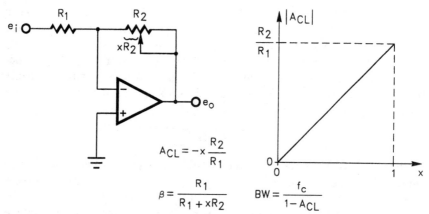

Figure 2.7 Adding a resistor to the previous inverting circuit produces linear control of variable gain and sets a defined gain maximum.

For more general applications, adding a resistor to the feedback network produces linear gain control and a defined upper limit for A_{CL}. Figure 2.7 illustrates this option for the inverting case with a fixed input resistor and a potentiometer controlling the primary feedback resistance. This configuration develops the linear gain control shown with a magnitude range from zero to R_2/R_1. Over this range, the gain follows a response described by the expression

$$A_{CL} = -x\frac{R_2}{R_1}$$

Once again, the application's minimum bandwidth and maximum error gain requirements dictate the maximum A_{CL}, as now determined by the feedback factor

$$\beta = \frac{R_1}{R_1 + xR_2}$$

This factor reproduces the same limits $|A_{CL}|_{max} = f_c/BW_{min} - 1$ and $|A_{CL}|_{max} = |G_e|_{max} - 1$ described earlier for the nonlinear inverting configuration.

Switching to the noninverting version of this linear gain control circuit produces similar but somewhat superior results. As shown in Fig. 2.8, this conversion to the noninverting configuration again offsets the A_{CL} response upward, and in this case delivers a gain range of 1 to $1 + R_2/R_1$. The feedback factor remains unchanged from the preceding inverting case, but the noninverting circuit's inherent bandwidth and error gain advantages extend the maximum practical gain limits to $A_{CL\,max} = f_c/BW_{min}$ and $A_{CL\,max} = G_{e\,max}$.

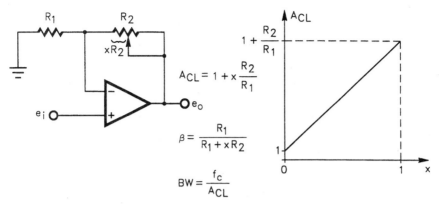

Figure 2.8 Switching to the noninverting version of Fig. 2.7 again produces a linear gain control response but sets a minimum gain of unity.

2.2.2 Variable difference gain

Adding variable gain to the difference amplifier requires careful attention to this amplifier's primary benefit, common-mode rejection. As described in Sec. 1.3, that rejection requires very precise resistance matching between its inverting and noninverting signal paths. Variable gain for this circuit could be implemented by simultaneously varying two of the basic circuit's four resistors. However, this straightforward option offers little hope of retaining the precise resistance matching required for high common-mode rejection. A simple alternative retains the matching through a differential tee network that varies gain by simultaneously altering the inverting and noninverting signal gains. While simple and suitable for many applications, this solution inherently produces a nonlinear gain control characteristic. No equivalent alternative produces linear gain control, but following the basic difference amplifier with either of the variable gain circuits presented in Figs. 2.7 and 2.8 serves applications that require that control characteristic.

Figure 2.9 shows the simple solution with the two R_2 resistors of the conventional difference amplifier divided into two equal parts and an R_2 potentiometer connected between the junctions of those parts. This configuration still requires the close resistor matching of the R_2/R_1 ratios of the fixed resistors but permits the addition of the R_2 potentiometer without disturbing the common-mode rejection produced by that matching. Adding the potentiometer shunts differential signal current between balanced impedance points of the inverting and noninverting sides of the circuit. Yet no common-mode current passes through the potentiometer. Any signal common to the e_1 and e_2 inputs produces equal signal components at the two ends of the potentiometer, avoiding disturbance of common-mode rejection. As a re-

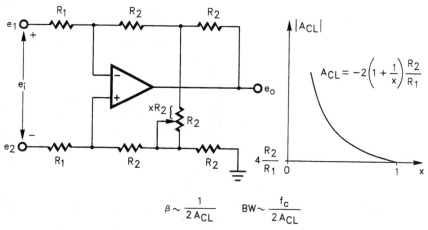

Figure 2.9 Forming a differential tee network adapts the difference amplifier to variable gain without disturbing the resistance matching critical to common-mode rejection.

sult, the signal current shunted by this tee connection produces a variable gain for the differential signal e_i equal to

$$A_{CL} = -2\left(1 + \frac{1}{x}\right)\frac{R_2}{R_1}$$

As shown in the figure, this nonlinear gain control response again delivers greater control sensitivity at lower gain levels.

The dual feedback paths introduced by the tee connection complicate the determination of β for bandwidth and error analyses. Output signal coupling through the potentiometer produces a component of positive feedback to the op amp's noninverting input. This positive feedback competes for control of the circuit with the conventional negative feedback delivered to the op amp's inverting input. Such a feedback combination could degrade stability except for the feedback attenuation introduced by the potentiometer. As long as the potentiometer setting fraction remains above zero, $x > 0$, the positive feedback path experiences a greater attenuation than the negative feedback path. Thus for $x > 0$, the negative feedback prevails and this condition encompasses the practical range of the A_{CL} response described.

The combination of the positive and negative feedback paths makes an exact expression for β more complex than useful. However, the approximation $\beta \approx 1/(2A_{CL})$ adequately predicts bandwidth and error gain performance. Then for a given application, a minimum bandwidth requirement BW_{min} combines with the classic bandwidth expression $BW = \beta f_c$ to set a maximum gain limit of

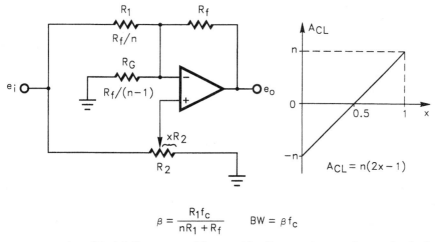

$$\beta = \frac{R_1 f_c}{n R_1 + R_f} \qquad BW = \beta f_c$$

Figure 2.10 A modified difference amplifier provides linear gain control spanning both gain polarities for a single input signal.

$$|A_{CL}|_{max} = \frac{f_c}{2BW_{min}}$$

Similarly, a maximum error gain requirement of $G_{e\ max}$ combines with $G_e = 1/\beta$ to produce the second limit equation

$$|A_{CL}|_{max} = \frac{G_{e\ max}}{2}$$

2.2.3 Bipolar variable gain

For a given input signal, the preceding circuits of this section produce variable gain magnitudes but restrict the associated gain polarities to either positive or negative. A surprisingly simple circuit configuration extends gain control to include both polarities. Shown in Fig. 2.10, this configuration[3] resembles a difference amplifier with the two normal inputs connected together. Through this connection, the input signal e_i receives both inverting and noninverting gains with the two gain magnitudes controlled by the setting of potentiometer R_2. At the setting extremes of $x = 0$ and $x = 1$, the circuit yields gain results derived by inspection. For $x = 0$, the noninverting input of the op amp connects directly to ground, shorting out the noninverting signal path. Then the circuit appears as a simple inverting amplifier with a gain of $A_{CL} = -R_f/R_1$ and making $R_1 = R_f/n$ produces $A_{CL} = -n$. For $x = 1$, signal e_i connects directly to the op amp's noninverting input and feedback transfers e_i to the op amp's inverting input as well. Then e_i appears at both ends of R_1, and this resistor conducts no signal current. However, in this condition, gain compensation resistor R_G supports the full e_i signal and the circuit performs as a noninverting amplifier having a gain of $A_{CL} = 1 + R_f/R_G$. Making $R_G = R_f/(n-1)$ reduces this gain to the noninverting complement of the inverting result, or $A_{CL} = +n$.

Between these setting extremes, the circuit supplies both inverting and noninverting gains as best defined with superposition analysis. In this analysis, connecting e_i to either the inverting or the noninverting input of the circuit and grounding the other input yields the corresponding gain. As described, superposition grounding of the R_2 input and driving the R_1 input with e_i yields an inverting gain of $A_{CLi} = -n$. Next, grounding the R_1 input and driving the R_2 input with e_i defines a noninverting gain of $A_{CLn} = 2xn$. Combining these two A_{CL} results yields the net gain $A_{CLi} + A_{CLn}$, or

$$A_{CL} = n(2x - 1)$$

Plotting this gain control response expression produces the linear control characteristic of the figure with a gain magnitude that varies from $-n$ to n and passes through zero at the potentiometer midpoint of $x = 0.5$.

Unlike preceding variable gain circuits, the feedback factor and the associated error gain and signal bandwidth remain constants for this bipolar control circuit. The feedback factor only depends on the fixed resistor network that couples the output signal to the op amp's inverting input. In this coupling, R_f, R_1, and R_G produce a voltage divider ratio that sets

$$\beta = \frac{R_1}{nR_1 + R_f}$$

This feedback factor defines the circuit's error gain $G_e = 1/\beta$ as

$$G_e = n + \frac{R_f}{R_1}$$

and defines $\text{BW} = \beta f_c$ as

$$\text{BW} = \beta f_c = \frac{f_c}{G_e}$$

Thus for the bipolar gain control circuit, the maximum gain magnitude n rather than the potentiometer fraction x controls the circuit's error gain and bandwidth.

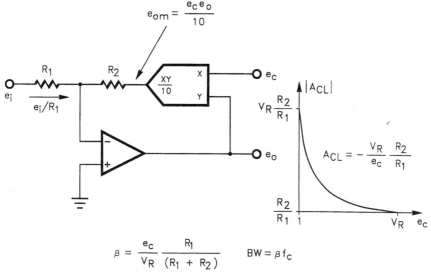

Figure 2.11   A feedback multiplier introduces electronic gain control in a continuously variable response defined by control voltage e_c.

2.2.4   Voltage-controlled gain

Potentiometer gain control, as developed by the preceding circuits, serves the requirements of manual control generally required for audio signal processing. However, more general variable-gain applications require electronic control for greater speed and automatic reaction to signal variations, such as in automatic gain control. For the latter applications, an analog multiplier replaces the potentiometer of the basic gain control circuits presented in Sec. 2.2.1. In Fig. 2.11, the inverting voltage-controlled alternative employs a fixed rather than variable R_2 resistance and uses a multiplier to vary the feedback signal supplied to this resistance. Through feedback, the control voltage e_c varies the output voltage e_o required to supply the feedback current conducted by R_2. In this way, the multiplier varies the circuit's effective feedback resistance, just as if a variable rather than fixed R_2 were in place. This solution inherently produces a nonlinear gain control characteristic, and no equivalent feedback control alternative produces a linear response. However, simply following a basic inverting or noninverting amplifier with a multiplier serves applications that require this linear control characteristic.

To quantify the performance characteristics of the inverting alternative presented here, consider the typical analog multiplier response function XY/V_R. There X and Y represent the signals applied to the multiplier's two inputs and V_R represents the reference voltage of the multiplier. For the circuit shown, $X = e_c$ and $Y = e_o$, making the multiplier output voltage $e_{om} = e_c e_o / V_R$. Feedback varies e_o, adjusting e_{om} to the value required to supply the feedback current initiated in R_1 and transferred to R_2. That current e_i / R_1 requires that $e_{om} = -(R_2/R_1)e_i$, and solving the two e_{om} expressions for e_o produces $e_o = -(V_R/e_c)$ (R_2/R_1). At one control extreme, $e_c = V_R$ for $e_o = -(R_2/R_1)e_i$, just as if a short circuit replaced the multiplier to make the circuit a simple inverting amplifier. Decreasing e_c from its maximum of V_R increases the e_o signal required by feedback, just as if the R_2 resistance had been increased. Thus the multiplier/R_2 combination produces an effective feedback resistance of $R'_2 = (V_R/e_c)R_2$. This effective resistance produces an inverting circuit gain of

$$A_{\mathrm{CL}} = -\frac{V_R}{e_c}\frac{R_2}{R_1}$$

As expressed, the control voltage e_c varies the effective resistance of the inverting amplifier's R_2 resistance in proportion to V_R/e_c. This produces the nonlinear $|A_{\mathrm{CL}}|$ versus e_c characteristic shown. The analogous noninverting configuration produces a gain of $A_{\mathrm{CL}} = 1 + (V_R/e_c)(R_2/R_1)$.

For the inverting case shown, feedback analysis again defines the error gain and bandwidth of this voltage-controlled circuit. With the multiplier, the effective voltage divider ratio of the feedback network becomes

$$\beta = \frac{e_c}{V_R}\frac{R_1}{R_1 + R_2}$$

This feedback factor varies with e_c, presenting a more complex frequency stability condition. In addition, the circuit encloses two active devices, the op amp and the multiplier, in a common feedback loop. One of the two devices must develop a dominant frequency response pole to establish a stable feedback condition. However, choosing a multiplier with a bandwidth much greater than the unity-gain crossover frequency of the op amp reduces stability considerations to the norm. Then the expression for β and the op amp's open-loop response define the circuit's frequency stability as described in another publication.[4] Given this feedback factor, $G_e = 1/\beta$ and $BW = \beta f_c$ define the error gain and bandwidth of the voltage-controlled, variable-gain, inverting amplifier.

For previous variable-gain circuits, the G_e and BW constraints generally defined the maximum practical gain. In this case, the multiplier introduces a new and often dominant limit. The offset, noise, and nonlinearity of the multiplier establish a minimum practical value for e_c. First the multiplier's input offset voltage V_{OSm} adds to e_c and potentially reverses the feedback polarity in the presence of small e_c signals. Including this offset in the feedback factor expression produces

$$\beta = \frac{e_c - V_{OSm}}{V_R} \frac{R_1}{R_1 + R_2}$$

Then any $e_c < V_{OSm}$ would reverse the polarity of β, producing positive rather than negative feedback and a latch condition. Also as e_c becomes small, the internal gain of the multiplier becomes large, amplifying the multiplier's input noise voltage to high levels. Finally, the nonlinearity of the multiplier response produces an error signal that can override the control of small e_c signals. Combined, these multiplier errors typically limit the circuit's practical gain control range to around 100:1.

2.3   Switched-Gain Amplifiers

Conceptually, switched gain with an op amp simply requires switching connections to the feedback network or to an input attenuator. However, in practice the switches introduce parasitic resistances and capacitances that potentially degrade gain accuracy, bandwidth, or frequency stability. This section evaluates these compromises for the various switching point alternatives of an op amp circuit. For high-speed electronic gain switching, switches at the op amp's summing junction produce the best alternative for gain accuracy, bandwidth, and cost.

2.3.1   Switching point alternatives

First, consider the four fundamental switching options illustrated in Fig. 2.12. There either SW_i, SW_o, SW_j, or SW_a switches gain by changing connections at a signal input, the op amp output, the summing junction, or an input attenuator. Considered individually, simple inspection reveals that each of these options switches the circuit's input-to-output gain supplied to an e_1 or e_2 input signal. Signal input switch SW_i, output switch SW_o, and summing-junction switch SW_j change the amplifier's feedback configuration to switch the gain supplied to either an e_1 or an e_2 input signal. Attenuator switch SW_a only changes the gain supplied to an e_2 signal. Other switch point options exist, but the four illustrated here serve to demonstrate the fundamental compromises of any switched-gain application of inverting, noninverting, or difference configurations of an op amp. In the figure, simple double-pole switches illustrate these options with two gain settings for each switch that connect either an a or b set of resistors. The switching methods shown can obviously be extended to virtually any number of gain settings through additional resistors and switch positions.

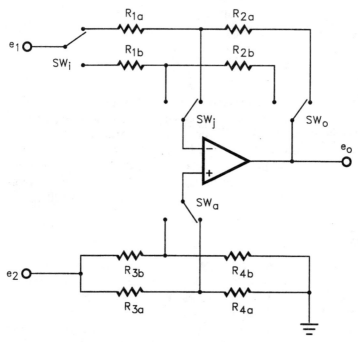

Figure 2.12 Gain switching with an op amp circuit can be performed at a signal input, at the op amp output, at the summing junction, or at an input attenuator.

For each switching option shown, performance compromises arise from the parasitic resistances and capacitances inherently introduced by the switches. Figure 2.13 models those parasitics for the a set of switch positions in Fig. 2.12. In this model, the various dashed-line R_S and C_S elements represent the switch parasitics. Depending on their locations, those parasitics potentially affect gain accuracy, bandwidth, or stability. First consider the gain accuracy disturbance produced by the presence of switches SW_i and SW_o, which reside in series with the circuit's feedback network. Signal input switch SW_i introduces R_{Si} and C_{Si}. Generally, the low output impedance of the e_1 signal source easily drives the C_{Si} capacitance, but the R_{Si} resistance appears in series with the feedback network, where this resistance degrades gain accuracy. Output switch SW_o produces similar compromises. This switch introduces R_{So} and C_{So} and the low impedance of the op amp output easily drives C_{So}, but the presence of R_{So} in the feedback path again produces a gain error. Compensating adjustments in feedback resistances can remove the initial gain errors produced by R_{Si} and R_{So}, but the resistances of electronic switches typically display high temperature coefficients, which would still preclude precise gain settings. For

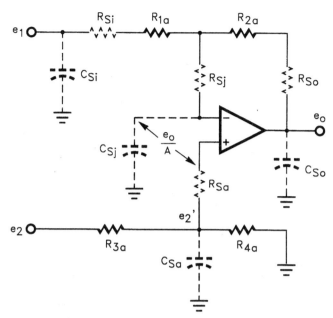

Figure 2.13 Parasitic resistances and capacitances introduced by the switches of Fig. 2.12 potentially degrade gain accuracy, bandwidth, or frequency stability, depending on the location of the parasitics.

this reason, the signal input switching and amplifier output switching options primarily serve lower speed requirements where relays provide low parasitic resistances. Alternately for these options, large MOS switches retain fairly low parasitic resistances and improve speed. However, both of these alternatives compromise cost.

Generally, a better solution to switch resistance errors results from placing the switches directly in series with the op amp's inputs, as with the SW_j and SW_a alternatives of Fig. 2.12. These switches introduce the resistances R_{Sj} and R_{Sa} in Fig. 2.13 in series with those inputs, and there the very low input signal currents produce virtually no gain error. However, the C_{Sa} and C_{Sj} capacitances restrict bandwidth and C_{Sj} potentially compromises frequency stability. Parasitic capacitance C_{Sa} appears at the op amp's noninverting input, where it bypasses the R_{4a} resistor, producing a bandwidth limit at $1/2\pi R_3 \| R_4 C_{Sa}$. This limit makes the switching of noninverting input signals a performance-limited alternative. Switching gains at the inverting input introduces a similar capacitance C_{Sj}, but that capacitance imposes a far lesser bandwidth limit. The op amp's virtual ground feature greatly reduces the signal shunting of C_{Sj} to avoid a similar bandwidth limit upon an e_1 input signal. Neglecting e_2 for this case, the op amp's virtual ground input reduces the signal impressed upon a C_{Sj} switch capacitance to e_o/A, where A is the high open-loop gain of the op amp.

Ideally, this amplifier action reduces the signal swing on this switch capacitance to virtually zero but the op amp gain has its own frequency limitation. As the frequency increases, gain A declines, making the e_o/A signal rise, and this ultimately sets a new but greater bandwidth limit for the circuit. Another treatment examines the bandwidth and the stability effects of capacitance at the op amp's inverting input in greater detail.[5] At more than any other point in the circuit, capacitance at that input can seriously degrade frequency stability due to a pole introduced in the feedback path. To restore stability, phase compensation counteracts the effect of this capacitance in a compromise that optimizes bandwidth. This phase compensation restricts bandwidth, but to a lesser degree than encountered with the previous noninverting input switching alternative.

2.3.2 Practical gain switching

The example of the preceding section demonstrates the gain-switching options that accommodate each of the fundamental configurations of an op amp voltage amplifier. However, for simplicity in that demonstration, the example uses far more resistors than necessary in practice. There, paralleled complete feedback networks and input attenuators quickly convey the concepts of gain switching for the analysis of parasitic error effects. However, in practice, series rather than parallel combinations of resistors produce the same gain-switching options with far fewer resistors. Further, the preceding example uses double-pole switches for the gain switching, again to simplify the discussion. In practice, the generally preferred electronic switches only produce single-pole action.

In a more practical example, Fig. 2.14 shows a switched-gain inverting amplifier having a series resistor combination and single-pole switches connected in series with the op amp's inverting input. As described, this combination optimizes gain accuracy, bandwidth, and cost for switched-gain applications. While shown with an inverting configuration, the practical modifications shown here readily adapt to the noninverting and difference amplifier connections as well. For those connections, simply replacing the potentiometers of the variable-gain circuits in Sec. 2.2 with the resistor network and switches shown produces analogous switched-gain alternatives. In the case illustrated, switches SW_1 through SW_n switch the circuit's closed-loop gain through multiple connections to a string of feedback resistors. In general, only one switch at a time makes a feedback connection and the resulting voltage gain follows from simple inverting amplifier

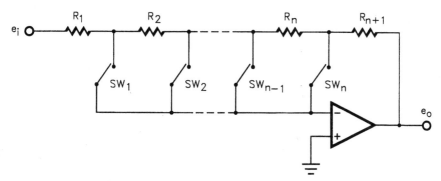

Figure 2.14 A more practical gain-switching example combines an inverting op amp with a series-connected feedback array and single-pole switches.

analysis. Then, switching the circuit's feedback network simply steps down the series string of resistors.

The option to make more than one switch active at a given time greatly expands the number of gain steps accessible. For example, activating SW_1 and SW_2 together shorts out R_2 while activating only SW_1 retains R_2 in the feedback network. With relatively few switches, the number of gain alternatives soon warrants Boolean design and analysis. However, activating multiple switches reintroduces feedback current in the switch resistances to produce gain errors.

References

1. J. Graeme, *Optimizing Op Amp Performance,* McGraw-Hill, New York, 1997.
2. D. Sheingold, "Calculating Resistances for Sum and Difference Networks," *Electronics,* June 12, 1975.
3. J. Graeme, "A Single Potentiometer Adjusts Op Amp's Gain over Bipolar Range," *Electronic Design,* p. 68, July 19, 1975.
4. Graeme, op cit., 1997.
5. Graeme, op cit., 1997.

3

Current Amplifiers

While universally known for processing signal voltages, the op amp also performs analogous tasks with signal currents. The equivalent current amplifiers accept an input signal current, process it, and deliver an output current signal. The straightforward processing of such signals would first convert the signal to a voltage, process the intermediate result in voltage mode with conventional circuitry, and then reconvert the resulting voltage to an output current. For applications requiring this three-step process the transimpedance and transconductance amplifiers of the next chapter perform the required conversions. However, many applications benefit from retaining the signal in the current mode to avoid the complexity and added errors introduced by these conversions. As described in this chapter, signal currents can be processed in one step by one op amp with performance characteristics that parallel those of voltage-mode counterparts. The key to their operation lies in a combination of negative and positive feedback that produces a current output in response to a current input. Examples that follow demonstrate this feedback control for noninverting, inverting, difference, integrating, and differentiating current amplifiers.

3.1 Basic Current Amplifiers[1]

Basic current amplifiers provide current-mode equivalents of the noninverting, inverting, and difference amplifiers of Chap. 1. Simple feedback connections again define circuit gains with design equations that parallel those of the voltage-mode equivalents. The noninverting current amplifier produces a positive current gain again determined by two feedback elements. The inverting counterpart produces negative current gain determined by analogous feedback elements. The difference current amplifier adds common-mode rejection to the circuit's performance. However, three distinctions separate these cur-

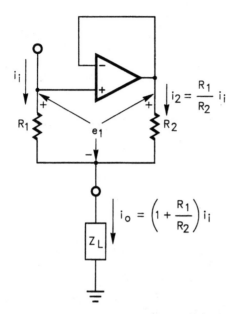

Figure 3.1 The noninverting current amplifier produces current gain by transferring the voltage developed on R_1 by i_i to a second resistor, R_2.

rent-mode amplifiers from the voltage-mode alternatives. First, the noninverting current amplifier permits gains of less than unity to form an attenuator rather than an amplifier. Next, signal connection requirements restrict the difference current amplifier to unity gain. Finally, all current-mode configurations present less than ideal input and output impedances. The impedances realized parallel those of the common-emitter transistor with load and source impedances reflected to the circuit's input and output scaled by the circuit's current gain.

3.1.1 Noninverting current amplifiers

Figure 3.1 shows the current-mode version of the noninverting amplifier as formed with a voltage follower and two gain-setting resistors. There, input current i_i develops a voltage e_1 on sense resistor R_1 and the follower reproduces this voltage on R_2. The resulting current in R_2, $i_2 = (R_1/R_2)i_i$, adds to i_i, and together the two currents drive load Z_L with an output current of

$$i_o = \left(1 + \frac{R_1}{R_2}\right) i_i$$

This output result parallels that of the voltage-mode noninverting amplifier, except for the inverted R_1/R_2 factor here. This inversion reflects the conversion from voltage to current gain, and now making R_2 smaller increases rather than decreases the circuit's gain. Note that

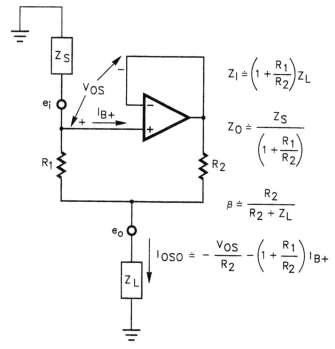

$$Z_I \doteq \left(1 + \frac{R_1}{R_2}\right)Z_L$$

$$Z_O \doteq \frac{Z_S}{\left(1 + \frac{R_1}{R_2}\right)}$$

$$\beta \doteq \frac{R_2}{R_2 + Z_L}$$

$$I_{OSO} \doteq -\frac{V_{OS}}{R_2} - \left(1 + \frac{R_1}{R_2}\right)I_{B+}$$

Figure 3.2 Modeling the Fig. 3.1 current amplifier defines the circuit's offset current, feedback factor, and terminal impedances.

the load Z_L shown here and in the following diagrams could be a series connection of several current conditioning circuits.

Other key performance characteristics of this circuit include its offset error, bandwidth, and terminal impedances. Figure 3.2 models the noninverting current amplifier for the related analyses and, for simplicity, this model neglects the i_i input signal but retains its Z_S source impedance. Beginning with offset error, the figure shows the op amp's input bias current I_{B+} and offset voltage V_{OS} that produce an output offset current I_{OSO}. Assuming Z_S to be large, essentially all of the I_{B+} current flows into the R_1 resistor, just like an input signal current. The circuit supplies its normal gain to this current, producing an output offset component equal to $-(1 + R_1/R_2)I_{B+}$. Adding to this component, V_{OS} produces a voltage on R_2 even in the absence of a voltage on R_1 for an output offset component $-V_{OS}/R_2$. Together, the two amplifier errors produce an output offset current of

$$I_{OSO} = -\frac{V_{OS}}{R_2} - \left(1 + \frac{R_1}{R_2}\right)I_{B+}$$

In conjunction with voltage swing requirements, this expression helps define the absolute values of R_1 and R_2 for a given application. Resistor R_1 resides between the source and the load, and any voltage developed on this resistor reduces the remaining voltage swing available for the load. This suggests making R_1 small; however, for a given gain this would also reduce R_2, increasing the first term of the I_{OSO} result. A compromise between these two effects guides the final resistor selection.

Feedback analysis of Fig. 3.2 defines the bandwidth of the noninverting current amplifier. For the typical resistive feedback case, an op amp circuit delivers a bandwidth $BW = \beta f_c$, where β is the circuit's feedback factor and f_c is the unity-gain crossover frequency of the op amp.[2] At first, the short-circuit feedback of the voltage follower here would suggest a feedback factor of unity, and indeed this feedback defines a negative feedback factor of $\beta_- = 1$. However, the circuit also includes a positive feedback path through the R_2 and R_1 connection between the op amp's output and its noninverting input. This added feedback converts the circuit's internal signal voltage into an output current. As always, the associated feedback factor equals the output-to-input voltage divider ratio of the feedback network. In this case, the network includes R_2, Z_L, R_1, and Z_S, but the typically large Z_S makes the feedback current negligible in the R_1, Z_S branch. Thus only R_2 and Z_L define a significant positive feedback signal and their divider ratio produces $\beta_+ = Z_L/(R_2 + Z_L)$. The circuit's negative and positive feedback factors combine to produce a net feedback factor[3] of $\beta = \beta_- - \beta_+$, or

$$\beta \cong \frac{R_2}{R_2 + Z_L}$$

The potentially reactive nature of Z_L precludes the use of the $BW = \beta f_c$ simplification that defines bandwidth for resistive feedback cases. However, the more typical case of a resistive load makes $Z_L = R_L$, restoring this bandwidth simplification, and there

$$\beta \cong \frac{R_2}{R_2 + R_L}, \qquad \text{for } Z_L = R_L$$

for

$$BW \cong \frac{R_2}{R_2 + R_L} f_c$$

The thus derived feedback factor also predicts numerous other performance characteristics, as analyzed in an earlier publication.[4] In particular, this factor permits the prediction of an op amp circuit's frequency stability through a combined plot of the circuit's $1/\beta$ curve and the op amp's open-loop response. In the general case, $1/\beta \cong 1 + Z_L/R_2$, and an inductive Z_L would produce a rising $1/\beta$ curve at higher frequencies. Such a curve could intercept the open-loop response curve with a high rate of closure, indicating response ringing, gain peaking, or even oscillation. In such cases, a capacitive bypass of Z_L would remove the $1/\beta$ rise to restore stability. However, too large of a bypass would restrict bandwidth unnecessarily, and the bypass selection should be guided by the stability analysis approach described in the publication referenced.

3.1.2 Terminal impedances of noninverting current amplifiers

The terminal impedances presented at the input and output of the noninverting current amplifier differ greatly from those of the voltage-mode counterpart. The voltage-mode circuit closely approximates the infinite input impedance and zero output impedance of an ideal voltage amplifier. For the current-mode equivalent, the ideal case would present zero input impedance and infinite output impedance. However, signal swings on the load and source compromise these impedances for the noninverting current amplifier and the results parallel the impedance transformations produced by a common-emitter transistor connection. Load impedances reflect to the input multiplied by the circuit's current gain and source impedances reflect to the output divided by that gain.

Further analysis of Fig. 3.2 defines these impedances beginning with the input impedance. Current supplied to the input terminal produces a nonzero voltage e_i, indicating nonzero input impedance. That voltage results from the flow of input current in R_1 and from the flow of the output current in Z_L. Normally, reserving most of the available voltage swing range for the load makes the effect of R_1 second order. Then, neglecting that effect and analyzing the current–voltage relationship of the e_i terminal defines the input impedance for the noninverting current amplifier as

$$Z_I \cong \left(1 + \frac{R_1}{R_2}\right)Z_L$$

Thus the input impedance equals the load impedance multiplied by the circuit's current gain, just like a common-emitter transistor.

A similar current–voltage relationship defines the circuit's output impedance. Current flow in the Z_L load develops a voltage e_o at the output terminal, and the circuit reflects this voltage to the source impedance Z_S. This develops an e_o/Z_S error current that flows into the circuit input to change the output current. Analysis of this effect defines the output impedance for the noninverting current amplifier as

$$Z_O \cong \frac{Z_S}{(1 + R_2/R_1)}$$

Thus the output impedance equals the source impedance divided by the circuit's current gain, again paralleling a common-emitter transistor. Both of the preceding impedances can be altered by the limitations of the op amp input impedance, open-loop gain, and common-mode rejection. However, the stated equations give sufficient approximations for these terminal impedances, except when approaching the frequency limit of the op amp as discussed later.

3.1.3   Noninverting current attenuators

The current-mode version of the noninverting amplifier also permits gain settings not attainable with the voltage-mode version. Specifically, the current-mode circuit permits gains of less than unity to form a current attenuator. This feature results from the bilateral nature of the circuit in which input and output reverse roles, as illustrated in Fig. 3.3. There the input current i_i drives the junction of the R_1, R_2 resistor

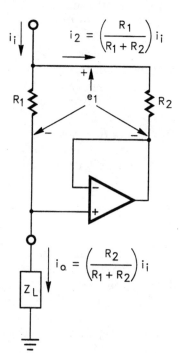

Figure 3.3 The bilateral nature of the noninverting current amplifier permits reversing the input and output connections to develop current gains less than unity.

connection and the circuit supplies only the R_1 current to the load Z_L. As before, the voltage follower establishes a voltage on R_2 equal to that developed on R_1 just as if the resistors were connected in parallel. Parallel connected resistors form a current divider, and the divider action defines the current supplied to R_1 and to the output as

$$i_o = \frac{R_2}{R_1 + R_2} i_i$$

The op amp output absorbs the remainder of the i_i current through the R_2 resistor.

The input–output switch of the noninverting current attenuator modifies the circuit's offset error, bandwidth, and terminal impedances, and Fig. 3.4 models the circuit for the related analyses. Beginning with offset error, the figure shows the op amp's input bias cur-

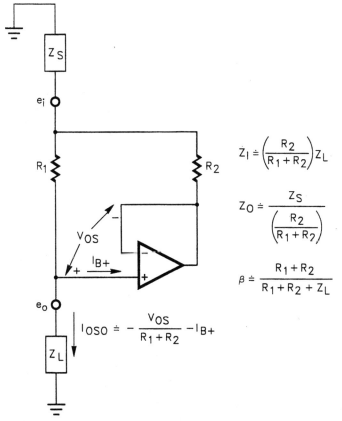

Figure 3.4 Modeling the Fig. 3.3 current attenuator defines the circuit's offset current, feedback factor, and terminal impedances.

rent I_{B+} and offset voltage V_{OS} that produce an output offset current I_{OSO}. To simplify the analysis, this model again neglects the i_i input signal and replaces that source with just its Z_S source impedance. Even in the absence of this signal, feedback impresses the voltage V_{OS} across the R_1, R_2 resistor network. Assuming Z_S to be large, this voltage develops an offset current of $-V_{OS}/(R_1 + R_2)$, and essentially all of this current flows into the Z_L load. Adding to this offset component, I_{B+} flows to the junction of R_1 and Z_L. Feedback controls the voltage across R_1 and R_2, fixing the current in R_1, and this prevents the flow of I_{B+} in that resistor. Thus I_{B+} must also flow into the Z_L load, and together the two amplifier offset errors produce an output offset current of

$$I_{OSO} = -\frac{V_{OS}}{R_1 + R_2} - I_{B+}$$

As before, this error expression helps guide the compromise selection of values for the R_1 and R_2 resistances. Making these values small increases I_{OSO}. However, making them large again reduces the input signal voltage swing transferable to the load.

Feedback analysis of Fig. 3.4 defines the bandwidth of the noninverting current attenuator. Once again, this circuit combines the unity negative feedback of the voltage follower with the positive feedback of the R_1, R_2 connection. The short-circuit negative feedback of the follower produces $\beta_- = 1$. The positive feedback couples through a network formed by R_2, Z_S, R_1, and Z_L, but the typically large Z_S makes its effect negligible. Then $\beta_+ = Z_L/(R_1 + R_2 + Z_L)$. The circuit's negative and positive feedback factors combine to produce a net feedback factor[5] of $\beta = \beta_- - \beta_+$, or

$$\beta \cong \frac{R_1 + R_2}{R_1 + R_2 + Z_L}$$

The feedback factor expressed here again warns of potential stability degradation through the corresponding $1/\beta$ response. In this case, $1/\beta \cong 1 + Z_L/(R_1 + R_2)$ and an inductive Z_L would produce a rising $1/\beta$ curve at higher frequencies. This can produce response ringing, gain peaking, or even oscillation, in which case stability restoration requires a capacitive bypass of Z_L.

As with the noninverting current amplifier, the potentially reactive nature of Z_L precludes the use of the BW $= \beta f_c$ simplification of resistive feedback cases. For the general case, this requires detailed feedback analysis to determine bandwidth.[6] However, the common case of a resistive load, $Z_L = R_L$, still permits the use of this simplification and then

$$\text{BW} \cong \frac{R_1 + R_2}{R_1 + R_2 + R_L} f_c, \quad \text{for } Z_L = R_L$$

Like the noninverting current amplifier before, the current attenuator here produces nonideal terminal impedances due to the presence of signal voltage swings on the source and load impedances. As before, the circuit's input and output impedances parallel the reflected impedance action of a common-emitter transistor. Specifically, the circuit's input impedance equals the load impedance multiplied by the circuit's current gain, and its output impedance equals the source impedance divided by that gain. Analysis of Fig. 3.4 shows these impedances to be approximated by

$$Z_I \cong \left(\frac{R_2}{R_1 + R_2} \right) Z_L$$

and

$$Z_O \cong \frac{Z_S}{R_2/(R_1 + R_2)}$$

Comparing these expressions with those of the preceding noninverting current amplifier shows that the attenuator decreases input impedance and increases output impedance to more closely approximate the ideal current amplifier.

3.1.4 Inverting current amplifiers

Figure 3.5 shows the current-mode version of this basic amplifier as formed with an op and two gain-setting feedback resistors. There input current i_i develops a voltage e_1 on sense resistor R_1 and the op amp feedback replicates this voltage on R_2 in order to maintain zero voltage between the op amp's inputs. The resulting current in R_2 drives the load Z_L with an output current of

$$i_o = -\frac{R_1}{R_2} i_i$$

This output result parallels that of the voltage-mode inverting amplifier, except for the inverted R_1/R_2 factor here. This inversion reflects the conversion from voltage to current gain, and now making R_2 smaller increases gain rather than decreasing it.

Figure 3.6 models the inverting current amplifier for analyses of its offset error, bandwidth, and terminal impedances. To simplify the analyses, this model again neglects the i_i input signal and replaces

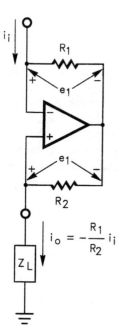

$$i_o = -\frac{R_1}{R_2} i_i$$

Figure 3.5 Like the noninverting equivalent, the inverting current amplifier produces current gain by transferring the voltage developed on R_1 by i_i to a second resistor, R_2.

that source with just its Z_S source impedance. Beginning with offset error, the figure shows the op amp's input bias currents I_{B^-} and I_{B^+} and offset voltage V_{OS} which produce an output offset current I_{OSO}. Superposition analysis guides the determination of the net offset result. First, superposition setting of V_{OS} and I_{B^+} to zero defines the offset contribution of I_{B^-}. Assuming Z_S to be large, essentially all of the I_{B^-} current flows into the R_1 resistor, just like an input signal current. Then, the circuit supplies its normal current gain to I_{B^-}, producing an output offset component equal to $(R_1/R_2)I_{B^-}$. Next, setting V_{OS} and I_{B^-} to zero defines the offset contribution of I_{B^+}. Under these conditions, feedback forces the voltage on R_2 to also be zero, and that resistor cannot accept the I_{B^+} current. Thus the I_{B^+} flows directly to the load, producing an output offset component of $-I_{B^+}$. Finally, setting I_{B^-} and I_{B^+} to zero defines the offset contribution of V_{OS}. Then, the op amp replicates this voltage on R_2, and this produces an output offset component equal to $-V_{OS}/R_2$. Together, the three amplifier errors produce an output offset current of

$$I_{OSO} = -\frac{V_{OS}}{R_2} + \frac{R_1}{R_2} I_{B^-} - I_{B^+}$$

As before, making R_2 small increases the offset contribution of V_{OS}. However, this resistor's selection no longer directly restricts the range

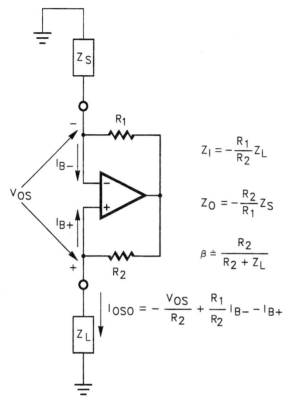

$$Z_I = -\frac{R_1}{R_2} Z_L$$

$$Z_O = -\frac{R_2}{R_1} Z_S$$

$$\beta \doteq \frac{R_2}{R_2 + Z_L}$$

$$I_{OSO} = -\frac{V_{OS}}{R_2} + \frac{R_1}{R_2} I_{B-} - I_{B+}$$

Figure 3.6 Modeling the Fig. 3.5 current amplifier defines the circuit's offset current, feedback factor, and
terminal impedances.

of the output voltage swing. Previously with the noninverting current
amplifier, the resistor selection produced a compromise between offset
and voltage swing performance because the R_1 resistor resided in series with the source and load. In this inverting case, the differential
inputs of the op amp reside in this series connection and feedback
maintains essentially zero voltage between those inputs. This removes the previous compromise and permits the selection of larger resistance values for reduced offsets. Now, the upper limit to resistance
levels results only from the output swing capability of the op amp,
where the voltage developed on the resistors requires the support of
the amplifier's output swing.

Further analysis of Fig. 3.6 defines the bandwidth of the inverting
current amplifier through its feedback factor. This circuit intentionally applies both positive and negative feedback, and their combined ef-

fects define a net feedback factor. R_1 and Z_S supply the negative feedback and produce a feedback factor $\beta_- = Z_S/(R_1 + Z_S)$. In practical cases, the input current source makes $Z_S \gg R_1$ for $\beta_- \cong 1$. R_2 and Z_L supply the positive feedback and produce a feedback factor $\beta_+ = Z_L/(R_2 + Z_L)$. Together the two feedback factors produce a net factor[7] for the circuit of $\beta = \beta_- - \beta_+$, or

$$\beta \cong \frac{R_2}{R_2 + Z_L}, \qquad \text{for } Z_S \gg R_1$$

As in the noninverting case, the feedback factor expressed here produces a rising $1/\beta$ curve in the presence of an inductive Z_L. This could compromise frequency stability and require capacitive bypass of Z_L, as described with that earlier case.

Combining the preceding β result with the resistive feedback relationship $\text{BW} = \beta f_c$ defines the inverting current amplifier's bandwidth for the most common case. There, a resistive load of $Z_L = R_L$ retains the feedback requirement for the $\text{BW} = \beta f_c$ simplification and yields

$$\text{BW} \cong \frac{R_2}{R_2 + R_L} f_c, \qquad \text{for } Z_S \gg R_1, \quad Z_L = R_L$$

where f_c is the unity-gain crossover frequency of the op amp. Bandwidth determination for the reactive Z_L case requires more detailed analysis using feedback response analysis.[8]

3.1.5 Terminal impedances of inverting current amplifiers

Finally, consider the terminal impedances presented at the input and output of this current amplifier. Further analysis of Fig. 3.6 produces results that again parallel the impedance transformations produced by a common-emitter transistor. Load impedances reflect to the input multiplied by the circuit's current gain and source impedances reflect to the output divided by that gain, as developed in the discussion of the noninverting case. For the inverting current amplifier this action produces

$$Z_I = -\frac{R_1}{R_2} Z_L$$

and

$$Z_O = -\frac{R_2}{R_1} Z_S$$

Here the polarity inversion of the circuit's current gain makes the input impedance negative to potentially produce oscillation. To avoid that, the net impedance in the input circuit must be kept positive through an impedance combination meeting the requirement

$$Z_S \geq \frac{R_1}{R_2} Z_L$$

where Z_S is the output impedance of the input signal source. From another perspective, this requirement assures that the circuit's negative feedback exceeds its positive feedback to retain stable control of the circuit.

The negative input impedance results from the circuit's inherent positive feedback, as required for the gain inversion, but it conflicts with a major application of current loop instrumentation. In remote monitoring the current mode reduces the significance of noise pickup on long lines. However, such lines also develop significant capacitances that shunt the source output to lower its net received impedance. To counteract the resulting oscillation potential, it may be necessary to add a decoupling resistor in series with the circuit input or a capacitive bypass of the load.

3.1.6 Difference current amplifiers

Two modifications to the inverting current amplifier convert it for difference operation, as shown in Fig. 3.7. There, a second input terminal connects to the op amp's noninverting input and equal-valued resistors balance the circuit. As before, the i_1 current develops a voltage on the upper resistor and op amp feedback replicates that voltage on the lower resistor. The resulting current in the lower one flows to the load as an output current component of $-i_1$. Since feedback controls the current in this resistor, the second input current i_2 cannot flow there and, instead, flows directly to the load. Together the two signal currents produce an output current of

$$i_o = i_2 - i_1$$

Because the current i_2 flows directly to the load, the circuit does not scale its effect and this restricts the circuit's current gain to unity.

Most of the performance characteristics of the difference current amplifier remain basically the same as described before for the inverting current amplifier. However, the equal-valued resistors here simplify the performance equations derived earlier. The difference configuration produces an output offset current of

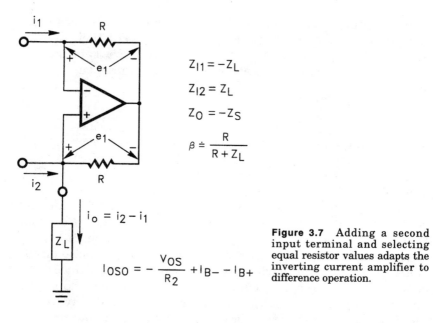

$$Z_{I1} = -Z_L$$

$$Z_{I2} = Z_L$$

$$Z_O = -Z_S$$

$$\beta \doteq \frac{R}{R + Z_L}$$

$i_o = i_2 - i_1$

$$I_{OSO} = -\frac{V_{OS}}{R_2} + I_{B-} - I_{B+}$$

Figure 3.7 Adding a second input terminal and selecting equal resistor values adapts the inverting current amplifier to difference operation.

$$I_{OSO} = -\frac{V_{OS}}{R} + I_{B^-} - I_{B^+}$$

a feedback factor of

$$\beta \cong \frac{R}{R + Z_L}, \qquad \text{for } Z_S \gg R$$

and a bandwidth for the $Z_L = R_L$ resistive feedback case of

$$\text{BW} \cong \frac{R}{R + R_L} f_c, \qquad \text{for } Z_S \gg R$$

With this circuit's equal-valued resistors, the input impedance for the i_1 input simplifies to

$$Z_{I1} = -Z_L$$

From before, this negative impedance requires that $Z_{S1} \geq Z_L$ to retain a net positive impedance in the input current loop for frequency stability. At the i_2 input, the signal current flows directly to the load, making

$$Z_{I2} = Z_L$$

The circuit's output impedance also follows directly from the inverting case and with the equal-valued resistors here becomes

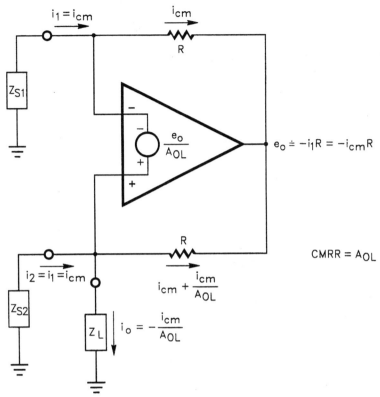

Figure 3.8 Modeling the Fig. 3.7 difference current amplifier shows that the op amp's open-loop gain determines the circuit's common-mode rejection.

$$Z_O = -Z_{S1}$$

The difference current amplifier introduces another key performance characteristic through common-mode rejection. Figure 3.8 models the circuit of Fig. 3.7 for analysis of this characteristic and shows the $i_2 = i_1 = i_{cm}$ condition to focus on the common-mode case. In the figure, the op amp's open-loop gain-error signal e_o/A_{OL} creates an output current error in the presence of these common-mode input currents. At first it might seem that the op amp's own common-mode rejection error should also contribute to the circuit's common-mode error. To examine this, first neglect the e_o/A_{OL} error with $A_{OL} = \infty$ to focus on the amplifier's common-mode condition. Then, common-mode input currents produce an output signal current of $i_o = i_2 - i_1 = 0$. This results in zero voltage across load Z_L for zero common-mode voltage at the op amp inputs. With no common-mode input voltage the op amp's common-mode rejection produces no error signal.

Only the finite A_{OL} gain of the op amp produces an output error current in response to i_{cm}. This error defines the circuit's common-mode rejection through the definition CMRR = A_D/A_{CM}, where A_D and A_{CM} represent the circuit's differential and common-mode gains. From before, the circuit produces $A_D = 1$ in its $i_o = i_2 - i_1$ response. To determine A_{CM}, consider the effect of the modeled input error source e_o/A_{OL}. The $i_1 = i_{cm}$ current develops an amplifier output voltage of $e_o = -i_{cm}R$ that reflects back to the circuit input as $e_o/A_{OL} = -i_{cm}R/A_{OL}$. The op amp impresses this input error voltage on the lower R resistor to develop an output error current of $i_o = i_{cm}/A_{OL}$. This reflects a common-mode current gain of $A_{CM} = i_o/i_{cm} = 1/A_{OL}$ and a common-mode rejection for the difference current amplifier of CMRR = A_D/A_{CM}, or

$$\text{CMRR} = A_{OL}$$

Previously, with the voltage-mode difference amplifier of Sec. 1.3, resistor mismatch also contributed to common-mode rejection error. With the current-mode case here, such mismatch produces only a negligible change in e_o, and that can be ignored in the e_o/A_{OL} signal that defines the preceding common-mode rejection.

3.2   Other Current-Mode Equivalents

Other common op amp functions adapt to the current mode following the feedback methods illustrated in the preceding sections. Switching a given application circuit from the voltage mode to the current mode simply requires employing the right combination of negative and positive feedback. This process generally results in a modified form of the preceding noninverting, inverting, or difference current amplifiers. To demonstrate the process, this section converts two other basic op amp functions, integration and differentiation, to the current mode. Both result from the addition of a capacitor to the inverting current amplifier. Other obvious conversions provide absolute-value conversion, peak detection, and clamping, as described in a previous publication.[9]

3.2.1   Integrating current amplifiers

Figure 3.9 shows the current-mode version of the basic integrator as formed with an op amp, an integrating capacitor, and two feedback resistors. Basically, the circuit forms an inverting current amplifier modified by the presence of the capacitor. With this modification, input current i_i develops a voltage e_1 on the R_1, C combination, and making R_1 large diverts the i_i current primarily to C over the circuit's useful frequency range. There, the e_1 developed represents the time

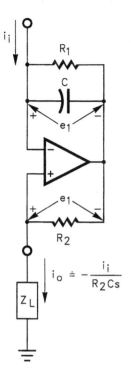

Figure 3.9 Adding capacitor C in parallel with R_1 converts an inverting current amplifier to a current integrator.

integral of i_i. Feedback replicates this voltage on R_2 in order to maintain zero voltage between the amplifier's inputs, and this converts the e_1 signal to an output current.

The resulting current in R_2 drives load Z_L with the current $i_o = -i_i/(R_2Cs + R_2/R_1)$. This result deviates from the ideal integrator response i_i/R_2Cs due to the circuit inclusion of R_1. However, R_1 must be included to maintain circuit stability. Without that resistor, the circuit's negative feedback factor becomes zero at dc, and the positive feedback of R_2 would dominate to produce a latch condition. Ultimately, the R_1 resistor imposes a limit on the circuit's lower frequency of useful operation. Still, a reasonable integrator response results for $R_2Cs > R_2/R_1$, where

$$i_o = -\frac{i_i}{R_2Cs}, \qquad \text{for } f > 1/2\pi R_1C$$

The frequency $f = 1/2\pi R_1C$ expressed imposes a lower-frequency 3-dB bandwidth limit for the integrator operation.

Figure 3.10 models the integrating current amplifier for analyses of its offset error, bandwidth, and terminal impedances. As before, this model neglects the i_i input signal and replaces that source with just

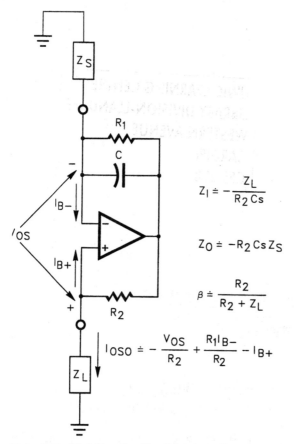

Figure 3.10 Modeling the Fig. 3.9 integrating current amplifier defines the circuit's offset, feedback factor, and terminal impedances.

its Z_S source impedance. Beginning with offset error, the figure shows the op amp's input bias currents I_{B-} and I_{B+} and offset voltage V_{OS}, which produce an output offset current I_{OSO}. Superposition analysis guides the determination of the net offset result. First, superposition setting V_{OS} and I_{B+} to zero defines the offset contribution of I_{B-}. In that case, assuming Z_S to be large forces essentially all of the I_{B-} current to flow into the R_1, C combination, just like an input signal current. At equilibrium, all of this dc current flows in R_1, producing an output offset component of $R_1 I_{B-}/R_2$. Next, setting V_{OS} and I_{B-} to zero reduces the voltage on R_1 to zero and defines the offset contribution of I_{B+}. Under these conditions, feedback forces the voltage on R_2 to also be zero, and that resistor cannot accept the I_{B+} current. Thus, I_{B+}

flows directly to the load, producing an output offset component of $-I_{B^+}$. Finally, setting I_{B^-} and I_{B^+} to zero defines the offset contribution of V_{OS}. Then, the op amp replicates this voltage on R_2 in order to retain zero voltage between its inputs, and this produces an output offset component equal to $-V_{OS}/R_2$. Together, the three amplifier errors produce an output offset current of

$$I_{OSO} = -\frac{V_{OS}}{R_2} + \frac{R_1 I_{B^-}}{R_2} - I_{B^+}$$

Further analysis of Fig. 3.10 defines the bandwidth of the integrating current amplifier through its feedback factor. This circuit intentionally applies both positive and negative feedback, and their combined effects define a net feedback factor. R_1, C, and Z_S produce a negative feedback factor of $\beta_- = Z_S/(Z_1 + Z_S)$, where $Z_1 = R_1/(1 + R_1Cs)$. In practical cases, the inclusion of R_1 in the circuit makes $Z_1 \ll Z_S$ for $\beta_- \cong 1$. R_2 and Z_L produce a positive feedback factor of $\beta_+ = Z_L/(R_2 + Z_L)$. Together, the two feedback factors produce a net feedback factor[10] for the circuit of $\beta = \beta_- - \beta_+$, or

$$\beta \cong \frac{R_2}{R_2 + Z_L}, \qquad \text{for } Z_1 \ll Z_S$$

Combining this result with the resistive feedback relationship BW $= \beta f_c$ defines the integrating current amplifier's bandwidth for the resistive load $Z_L = R_L$ case as

$$\text{BW} \cong \frac{R_2}{R_2 + R_L} f_c, \qquad \text{for } Z_1 \ll Z_S$$

where f_c is the unity-gain crossover frequency of the op amp. Normally, the BW $= \beta f_c$ relationship only applies to op amp configurations having purely resistive feedback and the integrating capacitor here inserts a reactive element in the negative feedback path. However, the $Z_1 \ll Z_S$ condition of this current amplifier makes $\beta_- \cong 1$ to avoid the reactive effects of the feedback at the bandwidth defining the intercept of $1/\beta$ and A_{OL}.

Further analysis of Fig. 3.10 defines the terminal impedances presented at the input and output of the integrating current amplifier. They again parallel the impedance transformations produced by a common-emitter transistor. As developed in Sec. 3.1.1, load impedances reflect to the input multiplied by the circuit's current gain and source impedances reflect to the output divided by that gain. For the integrating current amplifier this produces

$$Z_I \cong \frac{Z_L}{R_2 Cs}, \qquad \text{for } f > 1/2\pi R_1 C$$

and

$$Z_O \cong -R_2 Cs Z_S, \qquad \text{for } f > 1/2\pi R_1 C$$

3.2.2 Phase compensating the current integrator

The preceding approximations accurately reflect the circuit's terminal impedances over the circuit's useful frequency range, above $f = 1/2\pi R_1 C$. However, frequency stability evaluation requires a more exact expression for Z_I. As expressed, the inversion of the circuit's current gain makes the input impedance negative to potentially produce oscillation. To avoid that, the net impedance in the input circuit must be kept positive through selection of the R_1 resistance. From the beginning of this section, the circuit's exact current gain equals $A_i = i_o/i_i = -1/(R_2 Cs + R_2/R_1)$, and this makes the more accurate input impedance expression $Z_I = -Z_L/(R_2 Cs + R_2/R_1)$. Retaining a net positive impedance in the input circuit then requires that $Z_S > |Z_I| = Z_L/(R_2 Cs + R_2/R_1)$. This requirement reaches its greatest test at dc where the s factor goes to zero, reducing the requirement to $Z_S > R_1 Z_L/R_2$. Then, for stable operation R_1 must meet the condition

$$R_1 \le \frac{Z_S}{Z_L} R_2$$

where Z_S is the output impedance of the input signal source. From another perspective, this requirement assures that the circuit's negative feedback exceeds its positive feedback to retain stable control of the circuit. In practice, the very large Z_S values of input current sources make this requirement relatively easy to meet.

3.2.3 Differentiating current amplifiers

Figure 3.11 shows the current-mode version of the basic differentiator as formed with an op amp, a differentiating capacitor, and two feedback resistors. Basically, the circuit forms an inverting current amplifier modified by the presence of the capacitor. As before, input current i_i develops a voltage e_1 on R_1 and the op amp feedback replicates this voltage on the R_2, C combination. Making R_2 small impresses the majority of e_1 on C to produce a current proportional to the time derivative of i_i. The resulting current drives load Z_L with an output current of $i_o = -i_i R_1 Cs/(1 + R_2 Cs)$. This result deviates from the ideal $i_i R_1 Cs$

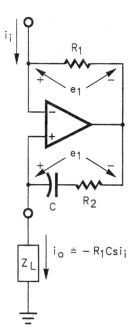

Figure 3.11 Adding capacitor C in series with R_2 converts an inverting current amplifier to a current differentiator.

differentiator response due to the circuit inclusion of R_2; however, this resistor must be included to maintain circuit stability. Without that resistor, the circuit's positive feedback factor would become unity at higher frequencies and dominate the ac response to produce oscillation. Ultimately, the R_2 resistor imposes a limit on the circuit's upper frequency of useful operation. Still, a reasonable differentiator response results for $R_2Cs < 1$, and there

$$i_o = -R_1Csi_i, \qquad \text{for } f < 1/2\pi R_2C$$

As expressed before, the frequency $f = 1/2\pi R_2C$ imposes a higher-frequency 3-dB limit to the circuit's useful bandwidth.

Figure 3.12 models the differentiating current amplifier for analyses of its offset error, bandwidth, and terminal impedances. As before, this model neglects the i_i input signal and replaces that source with just its Z_S source impedance. Beginning with offset error, the figure shows the op amp's input bias currents I_{B^-} and I_{B^+} and offset voltage V_{OS} that previously contributed to an I_{OSO} output offset current. For this case, superposition analysis again guides the determination of the net offset result. First, setting I_{B^+} to zero defines the offset contribution of V_{OS} and I_{B^-}. Under these conditions, V_{OS} and I_{B^-} produce dc circuit voltages blocked by the ac coupling effect of capacitor C and no output

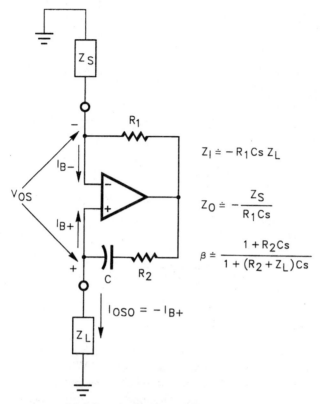

$$Z_I \doteq -R_1 C s \, Z_L$$

$$Z_O \doteq -\frac{Z_S}{R_1 C s}$$

$$\beta \doteq \frac{1 + R_2 C s}{1 + (R_2 + Z_L) C s}$$

Figure 3.12 Modeling the Fig. 3.11 differentiating current amplifier defines the circuit's offset, feedback factor, and terminal impedances.

current results. Feedback then holds the voltage across the R_2, C combination at a constant level of $-V_{OS} + I_B - R_1$ and no voltage change across capacitor C induces an output current. This same restriction of current in R_2 and C leaves only one path for I_{B+}, and that current flows directly to the load, producing an output offset current of

$$I_{OSO} = -I_{B+}$$

Next, consider the terminal impedances presented at the input and output of this circuit. Like the inverting current amplifier, load impedances reflect to the input multiplied by the circuit's current gain and source impedances reflect to the output divided by that gain. For the differentiating current amplifier this produces

$$Z_I \cong -R_1 Cs Z_L, \quad \text{for } f < 1/2\pi R_2 C$$

and

$$Z_O \cong -\frac{Z_S}{R_1 Cs}, \quad \text{for } f < 1/2\pi R_2 C$$

Further analysis of Fig. 3.12 defines the bandwidth of the differentiating current amplifier through its feedback factor. As with preceding configurations, this circuit intentionally applies both positive and negative feedback and their combined effects define a net feedback factor. R_1 and Z_S produce a negative feedback factor of $\beta_- = Z_S/(R_1 + Z_S)$, and practical cases make $Z_S \gg R_1$ for $\beta_- \cong 1$. R_2, C, and Z_L produce a positive feedback factor of $\beta_+ = Z_L Cs/[1 + (R_2 + Z_L)Cs]$. Then the circuit's two feedback factors produce a net factor[11] for the circuit of $\beta = \beta_- - \beta_+$, or

$$\beta \cong \frac{1 + R_2 Cs}{1 + (R_2 + Z_L)Cs}, \quad \text{for } R_1 \ll Z_S$$

The reactive nature of this feedback factor clearly precludes the use of the BW $= \beta f_c$ approximation that serves resistive feedback cases. Bandwidth determination for the general differentiating current amplifier requires a more detailed analysis as facilitated by feedback response plots. Plotting an op amp circuit's $1/\beta$ response along with the amplifier's open-loop gain magnitude then provides graphical predictions of bandwidth and frequency stability.[12] The arbitrary Z_L included in the β expression further precludes the determination of a general solution, but for many cases a resistive load converts Z_L to just R_L. Then, the differentiating current amplifier produces a $1/\beta$ response expressed by

$$\frac{1}{\beta} \cong \frac{1 + (R_2 + R_L)Cs}{1 + R_2 Cs}, \quad \text{for } R_1 \ll Z_S, \quad Z_L = R_L$$

Figure 3.13 shows the $1/\beta$ plot of this response along with the A_{OL} open-loop gain of a typical op amp. The latter response crosses the 0-dB gain axis at the unity-gain crossover frequency f_c. As expressed before, the $1/\beta$ curve begins a rise due to a response zero at $f_z = 1/2\pi(R_2 + R_L)C$. This rise potentially intercepts the A_{OL} response at f_i with a 40-dB per decade rate of closure or slope difference, and that would indicate oscillation. However, the presence of R_2 produces a pole in the $1/\beta$ response at $f_p = 1/2\pi R_2 C$ to terminate that rise and preserve stability.

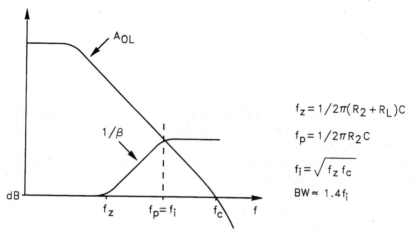

$$f_z = 1/2\pi(R_2 + R_L)C$$

$$f_p = 1/2\pi R_2 C$$

$$f_i = \sqrt{f_z f_c}$$

$$BW \approx 1.4 f_i$$

Figure 3.13 Graphical analysis of the Fig. 3.11 differentiating current amplifier defines the circuit's stability and bandwidth constraints through the associated A_{OL} and $1/\beta$ responses.

3.2.4 Phase compensating the current differentiator

Stability concerns would suggest making R_2 large to move the resulting f_p well below the f_i intercept. However, this choice restricts bandwidth unnecessarily. At higher frequencies, the impedance of R_2 exceeds that of C and the circuit reverts to an inverting current amplifier. Thus the break frequency of R_2 and C defines a potential bandwidth limit for the circuit. Similarly, the crossing of the circuit's $1/\beta$ and A_{OL} curves in the figure defines a second potential bandwidth limit. At the f_i intercept of these two curves, the feedback demand for gain $1/\beta$ reaches the level of the available gain A_{OL}, and any further increase in signal frequency results in response roll off. In compromise, setting the two bandwidth limits at the same frequency optimizes bandwidth. Then, $f_p = f_i$, and this results in 45° of phase margin for the optimal stability-versus-bandwidth compromise.

Designing for this compromise requires the knowledge of intercept frequency f_i, as determined by graphical analysis. In Fig. 3.13, the intersecting curves define a triangle with its peak at f_i. A single zero defines the left side of the triangle to produce a 20-dB per decade rise beginning at f_z. Similarly, a single pole from the op amp roll off defines the right side of the triangle, producing a −20-dB per decade decline, ending at f_c. Thus equal and opposite slopes make this an isosceles triangle and the f_i peak resides at the midpoint between f_z and f_c. Given the logarithmic nature of the frequency scale, the mathematical expression of this midpoint becomes $\log(f_i) = [\log(f_z) + \log(f_c)]/2$. Solving for f_i shows it to be the geometric mean of f_z and f_c, or

$$f_i = \sqrt{f_z f_c}, \qquad \text{when } f_p = f_i$$

The 45° phase margin of the $f_p = f_i$ condition introduces a 3-dB gain peak, which increases the -3-dB bandwidth limit to around BW = $1.4f_i$.

Selecting the R_2 phase compensation resistor for the $f_p = f_i$ condition requires a design equation, as derived from previous expressions. From before, $f_p = f_i = 1/2\pi R_2 C$, $f_z = 1/2\pi (R_2 + R_L)C$, and $f_i = \sqrt{f_z f_c}$. Directly combining these equations and solving for R_2 produces a complex equation that offers little design insight. However, a practical design constraint simplifies the analysis. In Fig. 3.12, both R_2 and R_L conduct the same current and develop voltages supported by the op amp's output voltage range. To allocate most of that output range to R_L, practical circuits make $R_2 \ll R_L$, and this produces $f_z \cong 1/2\pi R_L C$. Then, the solution for $f_p = f_i$ makes

$$R_2 = \sqrt{\frac{R_L}{2\pi C f_c}}, \qquad \text{for } R_2 \ll R_L$$

References

1. J. Graeme, "Manipulate Current Signals with Op Amps," *EDN*, p. 147, August 8, 1985.
2. J. Graeme, *Optimizing Op Amp Performance*, McGraw-Hill, New York, 1997.
3. Graeme, op cit., 1997.
4. Graeme, op cit., 1997.
5. Graeme, op cit., 1997.
6. Graeme, op cit., 1997.
7. Graeme, op cit., 1997.
8. Graeme, op cit., 1997.
9. Graeme, op cit., 1985.
10. Graeme, op cit., 1997.
11. Graeme, op cit., 1997.
12. Graeme, op cit., 1997.

4

Transimpedance and Transconductance Amplifiers

The preceding chapters present voltage and current amplifiers where the input and output signals share the same voltage or current form. This chapter develops amplifiers that provide conversions between the two forms while adding amplification. Transimpedance amplifiers convert a current input into a voltage output and transconductance amplifiers provide the converse role of converting an input voltage into an output current. As such, these amplifiers display gains expressed by an impedance or a conductance. Both of these amplifier types depart from the more common amplifier applications of op amps, but the traditional analysis methods still apply in the determination of circuit performance.

One op amp configuration emerges as the de facto standard for the transimpedance amplifier role due to its circuit simplicity and the near ideal terminal impedances it presents. The current-to-voltage configuration consists of just an op amp and a feedback resistor and presents near zero impedances at both its input and output terminals. However, this simple configuration often displays a surprisingly complex ac behavior that complicates the determination of bandwidth, requires the addition of phase compensation, and makes the noise response frequency-dependent. Analyses reduce these complexities to simple design equations that predict performance and prescribe phase compensation.

Modified difference amplifiers serve as the basis for transconductance amplifiers through positive feedback connections that produce the voltage-to-current conversion. The multielement feedback of the basic configuration obscures intuitive anticipation of circuit performance but feedback analysis defines simple results. Requirements for precise resistance matching limit the economic practicality of this basic transconductance amplifier and encourage the use of a boot-strapped alternative. The latter employs pretrimmed difference amplifiers to ensure the required matching but requires the addition of a bootstrap voltage follower. In a side benefit, the impedance isolation of the follower restores intuitive insight into the circuit's performance. For both transconductance configurations, analysis yields simple equations for the terminal impedances that express the fundamental quality of the voltage-to-current conversion.

4.1   Transimpedance Amplifiers[1]

As its name suggests, this amplifier converts an input current into an output voltage, giving rise to the "current-to-voltage converter" alias. Simple in structure, the transimpedance amplifier consists of just an op amp and a feedback resistor. This combination closely approximates the zero input and output impedances of the ideal current-to-voltage converter. However, high-gain applications of this circuit encounter more complex stability, bandwidth, and noise constraints. The simple resistor feedback would suggest that the circuit produces a bandwidth defined by the common op amp approximation of $BW = \beta f_c$. Often, however, this approximation applies only to resistive feedback cases, and even then, capacitances at this circuit's input potentially make the net feedback reactive. Further, the $BW = \beta f_c$ approximation assumes that parasitic capacitances shunting the feedback resistor do not introduce a bandwidth limit independently. The combination of these possibilities defines three potential bandwidth limits for the transimpedance amplifier, and each must be evaluated for a given application. Reactive feedback cases also threaten circuit stability and may require phase compensation. When required, simple capacitive bypass of the feedback resistor provides this compensation, as guided by design equations.

4.1.1   Current-to-voltage converters

Figure 4.1 shows the op amp and feedback resistor of the fundamental transimpedance amplifier. Also shown in dashed lines, feedback capacitor C_f provides phase compensation for high-gain applications as described later. Ideally a transimpedance amplifier would present zero impedance at both its input and its output and the circuit shown closely approximates this case. As with all op amp circuits, loop gain reduces the output impedance of the circuit to essentially zero over most of the circuit's useful frequency range. This circuit also approximates the zero input impedance condition through the virtual ground of its feedback input. Ideally, feedback forces the transimpedance am-

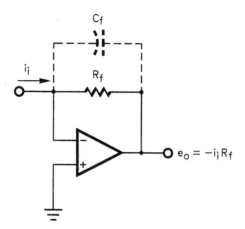

$e_o = -i_i R_f$

Figure 4.1 The transimpedance amplifier requires just an op amp and a feedback resistor to convert an input current to an output voltage.

plifier to accept the i_i input current without producing any voltage at the circuit input. In this feedback relationship i_i initially reacts with the high impedance of the op amp input, and the amplifier's open-loop gain amplifies the resulting input voltage, driving the feedback resistor R_f with output voltage e_o. This drive diverts i_i away from the amplifier's input impedance to R_f, restoring virtually zero voltage at the op amp input. In the process, the amplifier develops an output voltage equal to the input current times the feedback resistance for

$$e_o = -i_i R_f$$

In practice, the transimpedance amplifier presents a small but nonzero input impedance due to the finite open-loop gain of the op amp. The model of Fig. 4.2 supports the analysis of this impedance through the e_o/A_{OL} error source shown between the amplifier inputs. This model includes the optional phase compensation capacitor C_f to produce a net feedback impedance Z_f. However, for applications not requiring C_f, Z_f reverts to the basic R_f. For the case shown, input current i_i produces an output voltage of $e_o = -i_i Z_f$ and the op amp requires a differential input voltage of $e_o/A_{OL} = i_i Z_f/A_{OL}$ to support this output. By definition, input impedance equals the voltage developed by an input current divided by that current. In this case, i_i represents the current and e_o/A_{OL} represents the voltage for $Z_I = (e_o/A_{OL})/i_i$, or

$$Z_I = \frac{Z_f}{A_{OL}}$$

Thus the input impedance of the transimpedance amplifier equals the feedback impedance divided by the open-loop gain of the op amp.

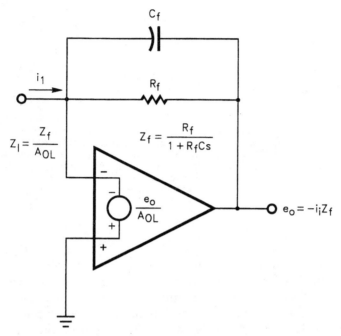

Figure 4.2 Op amp's finite open-loop gain results in a nonzero input impedance for the transimpedance amplifier.

At lower frequencies, A_{OL} remains large and Z_I closely approximates the ideal zero input impedance. At higher frequencies, Z_I rises due to the A_{OL} roll off, and this impedance can appear inductive. Combined with a capacitive source impedance, such an input impedance potentially produces a resonant circuit,[2] and this gives rise to the need for the C_f phase compensation, as defined in a later analysis.

Error analysis of the transimpedance amplifier continues with the model of Fig. 4.3. There the op amp's V_{OS} and I_{B-} input errors develop an output offset and the R_S and C_S components of the signal source impedance contribute to the circuit's feedback factor. Also, this model shows the C_{ia} input capacitance of the op amp removed from the amplifier for the feedback analysis. For the transimpedance amplifier, this capacitance includes both differential and common-mode input capacitances of the op amp and $C_{ia} = C_{id} + C_{icm}$.

Superposition guides the offset analysis considering one offset source at a time. First, superposition setting V_{OS} to zero defines the offset effect of I_{B-}. Typically, the large R_S of a signal current source diverts the I_{B-} current to R_f, where the current develops an output offset component equal to $R_f I_{B-}$. Next, setting I_{B-} to zero defines the effect of the op amp's input offset voltage V_{OS}. Assuming R_S to be large,

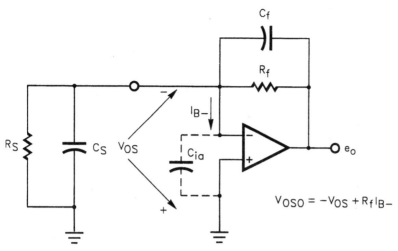

Figure 4.3 For the transimpedance amplifier of Fig. 4.1, amplifier offset errors define an output offset and feedback elements define the circuit's feedback factor.

the circuit supplies unity gain to this offset for an output offset component equal to $-V_{OS}$. Together, the two amplifier errors produce an output offset voltage of

$$V_{OSO} = -V_{OS} + R_f I_{B^-}$$

In high-gain applications of the transimpedance amplifier, R_f becomes large and the $R_f I_{B^-}$ term dominates the output offset. To reduce this offset component, the addition of the traditional compensation resistor R_C in Fig. 4.4 produces a counteracting effect. There, the amplifier's other input bias current I_{B^+} develops the voltage $-R_C I_{B^+}$ on the added compensation resistor. Assuming a large source impedance, the circuit transfers this voltage to the circuit output with unity gain, modifying the output offset equation to

$$V_{OSO} = -V_{OS} + R_f I_{B^-} - R_C I_{B^+}$$

This expression reflects a reduced V_{OSO} through two matching characteristics of the compensation approach. First, the matched inputs of the op amp make $I_{B^+} \approx I_{B^-}$. Then, making $R_C = R_f$ matches the voltage drops produced by the two currents to reduce the output offset to $V_{OSO} \approx V_{OS}$. Adding bypass capacitor C_B around R_C removes the higher-frequency component of noise introduced by this compensation resistor. For this high-gain case, the parasitic capacitance C_p may produce a dominant bandwidth limit, as examined next.

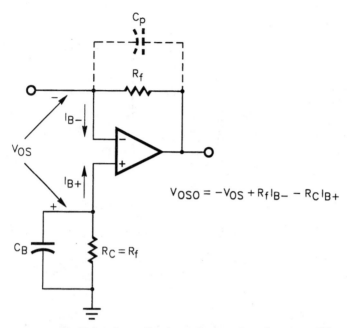

Figure 4.4 In high-gain applications, the transimpedance amplifier benefits from the inclusion of offset compensation resistor R_C and may be bandwidth limited by parasitic capacitance C_p.

4.1.2   Transimpedance amplifier bandwidth

For the typical resistive-feedback op amp circuit a frequency-independent feedback factor defines signal bandwidth through the relationship[3] $BW = \beta f_c$. The transimpedance amplifier uses simple resistive feedback that would suggest the application of this relationship. However, the transimpedance amplifier feedback remains vulnerable to capacitance in the input circuit that makes the feedback factor frequency-dependent, potentially invalidating the $BW = \beta f_c$ relationship. Also, high-gain applications of the circuit make the R_f feedback resistance vulnerable to the capacitance shunting of parasitic capacitance, potentially introducing an independent bandwidth limit.

Three cases illustrate the potential bandwidth results as determined by low-gain, high-gain, and intermediate-gain applications. Determining which case applies for a given application requires the evaluation of the bandwidth results predicted by all three. Then, the lowest bandwidth prediction defines the applicable case. Figure 4.5 models the transimpedance amplifier of Fig. 4.3 for bandwidth analysis. For simplicity, this model neglects the R_C compensation resistor of Fig. 4.4 since the C_B bypass of that resistor removes its effect well before the frequency of the bandwidth limit. In Fig. 4.5, assuming the

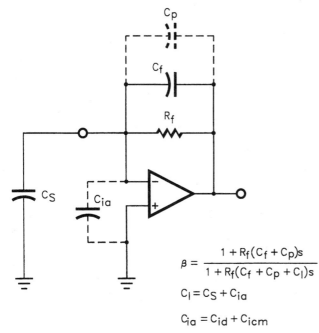

$$\beta = \frac{1 + R_f(C_f + C_p)s}{1 + R_f(C_f + C_p + C_l)s}$$

$$C_l = C_S + C_{ia}$$

$$C_{ia} = C_{id} + C_{icm}$$

Figure 4.5 Modeling the transimpedance amplifier of Fig. 4.3 defines the circuit's feedback factor for bandwidth and stability analyses.

previous R_S to be large permits neglecting that resistance, and including both the C_f and C_p feedback capacitances accommodates all cases. For high-gain applications, R_f becomes large and vulnerable to the signal shunting of the inevitable parasitic capacitance C_p. This case requires no phase compensation and $C_f = 0$. Then, simple RC roll off defines the circuit's bandwidth as BW $= 1/2\pi R_f C_p$, where C_p is the parasitic capacitance shunting R_f.

Other gain cases require the evaluation of the circuit's feedback factor, and numerous elements of the figure contribute to the output-to-input voltage-divider ratio that defines this factor. R_f, C_f, C_p, C_S, and C_{ia} form the feedback voltage divider and produce a feedback factor of

$$\beta = \frac{1 + R_f(C_f + C_p)s}{1 + R_f(C_f + C_p + C_I)s}$$

where $C_I = C_S + C_{ia}$ and $C_{ia} = C_{id} + C_{icm}$. Here, C_{id} represents the op amp's differential input capacitance and C_{icm} represents its common-mode counterpart. Depending on the relative magnitudes of R_f, C_f, and C_I, this feedback factor may or may not produce a frequency-dependent feedback factor within the useful frequency range of the op amp. For low-gain applications, R_f remains small and the preceding expression reduces to the unity feedback condition of $\beta = 1$ within this useful range for BW $= \beta f_c = f_c$.

Between the high- and low-gain extremes described, a more complex feedback relationship defines the circuit's bandwidth. In such cases, the C_f feedback phase compensation capacitor must be added to preserve frequency stability, as guided by the circuit's $1/\beta$ response. From the expression for β, the transimpedance amplifier produces a $1/\beta$ response described by

$$\frac{1}{\beta} = \frac{1 + R_f(C_f + C_p + C_I)s}{1 + R_f(C_f + C_p)s}$$

where $C_I = C_S + C_{ia}$ and $C_{ia} = C_{id} + C_{icm}$. Figure 4.6 shows the associated $1/\beta$ plot with the A_{OL} open-loop gain response of a typical op amp and the resulting I–V gain. As expressed before, the $1/\beta$ curve begins a rise due to a response zero at $f_z = 1/2\pi R_f(C_f + C_p + C_I)$. This rise potentially intercepts the A_{OL} response at f_i with a 40-dB per decade rate of closure or slope difference, and that would indicate oscillation.[4] However, the presence of C_f and C_p produces a pole in the $1/\beta$ response at $f_p = 1/2\pi R_f(C_f + C_p)$ to terminate that rise and preserve stability.

Stability concerns would suggest making C_f large to move the resulting f_p well below the f_i intercept. However, this choice unnecessarily restricts bandwidth. At higher frequencies, C_f and C_p bypass the R_f feedback resistor that converts an input current into an output voltage. Thus the break frequency of $C_f + C_p$ with R_f defines a potential bandwidth limit for the circuit. Similarly, the crossing of the $1/\beta$ and A_{OL} curves defines a further potential bandwidth limit. At the f_i intercept of these two curves, the increasing feedback demand for gain $1/\beta$ equals the available gain A_{OL}, and any further increase in signal frequency results in response roll off. In compromise, setting the present two bandwidth limits at the same frequency optimizes the potential bandwidth result. Then $f_p = f_i$, and this produces a phase margin of $\phi_m = 45°$ for a good stability-versus-bandwidth compromise.

Designing for this compromise requires the knowledge of the intercept frequency f_i and graphical analysis defines it in terms of known quantities. In Fig. 4.6, the intersecting $1/\beta$ and A_{OL} curves produce a triangle with its peak at f_i. A single zero defines the left side of the triangle, developing a 20-dB per decade rising slope beginning at f_z. Similarly, a single pole from the op amp roll off defines the right side of the triangle with a -20-dB per decade declining slope ending at f_c. Thus equal and opposite slopes make this an isosceles triangle and the f_i peak resides at the midpoint between f_z and f_c. Given the loga-

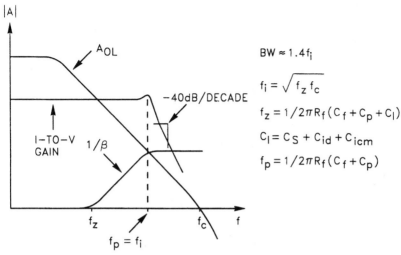

$|A|$

A_{OL}

$-40dB/DECADE$

I-TO-V
GAIN $1/\beta$

f_z

$f_p = f_i$

f_c

f

$BW \approx 1.4 f_i$

$f_i = \sqrt{f_z f_c}$

$f_z = 1/2\pi R_f (C_f + C_p + C_I)$

$C_I = C_S + C_{id} + C_{icm}$

$f_p = 1/2\pi R_f (C_f + C_p)$

Figure 4.6 Graphical feedback analysis defines an optimum bandwidth result that places the f_p pole of the $1/\beta$ response at the f_i intercept frequency of $1/\beta$ and A_{OL}.

rithmic nature of the frequency scale, the mathematical expression of this midpoint becomes $\log(f_i) = [\log(f_z) + \log(f_c)]/2$. Solving for f_i shows it to be the geometric mean of f_z and f_c, or

$$f_i = \sqrt{f_z f_c}$$

Then, the 45° phase margin produced by making $f_p = f_i$ introduces a 3-dB gain peak that increases the -3-dB bandwidth limit to around

$$BW \cong 1.4 f_i$$

However, f_i remains an unknown until the selection of the phase compensating C_f defines f_z, as described next.

4.1.3 Transimpedance amplifier phase compensation

Designing for the $f_p = f_i$ phase compensation condition requires an equation for the C_f selection, as derived here from previous expressions. From before, $f_p = 1/2\pi R_f(C_f + C_p)$ and $f_z = 1/2\pi R_f(C_f + C_p + C_I)$. Combining these equations for the previous $f_p = f_i$ and $f_i = \sqrt{f_z f_c}$ conditions produces a complex equation that offers little design insight. However, defining a fictitious capacitance $C_c = 1/2\pi R_f f_c$ reduces this result to a more intuitive form. Here, C_c represents that value of capacitance that would break with R_f at the op amp's unity-gain crossover frequency f_c. Using that value reduces the C_f design equation to

$$C_f = \frac{C_c}{2}\left(1 + \sqrt{1 + 4\frac{C_I}{C_c}}\right) - C_p$$

for $\phi_m = 45°$ and where $C_c = 1/2\pi R_f f_c$, $C_I = C_S + C_{ia}$, and $C_{ia} = C_{id} + C_{icm}$. As this equation expresses, the presence of parasitic capacitance C_p reduces the C_f bypass of R_f required to preserve stability. Because the equation subtracts C_p, this equation will sometimes deliver negative numbers for C_f. Those cases indicate that the specific application analyzed requires no phase compensation beyond that automatically provided by C_p.

A simplification of this result often applies for a common application of the transimpedance amplifiers, photodiode monitoring.[5] In these applications, large photodiodes often make $C_S \gg C_{ia} + C_f + C_p$ and require a phase compensation $C_f \gg C_p$. Under these conditions, the key equations defining the circuit's $1/\beta$ response reduce to $f_z \approx 1/2\pi R_f C_S$ and $f_p \approx 1/2\pi R_f C_f$. Then applying the $f_p = f_i$ and $f_i = \sqrt{f_z f_c}$ conditions of the phase compensation described produces the simplified design equation

$$C_f = \sqrt{\frac{C_S}{2\pi R_f f_c}}$$

when $C_S \gg C_{ia} + C_f + C_p$ and $C_f \gg C_p$.

Given C_f for a specific application, the circuit's bandwidth follows from previous results. As described before, low-gain applications do not require the C_f phase compensation and simple resistive feedback controls bandwidth for BW $= \beta f_c = f_c$. Similarly, high-gain applications do not require C_f because the parasitic capacitance C_p automatically fills that roll and makes BW $= 1/2\pi R_f C_p$. Only the intermediate gain cases require this phase compensation and then, from before,

$$BW \cong 1.4\sqrt{f_z f_c}$$

where $f_z = 1/2\pi R_f(C_f + C_p + C_I)$ and $C_I = C_S + C_{id} + C_{icm}$. To determine the bandwidth for a given application, all three potential bandwidth limits must be evaluated. The lowest result of the three defines the actual bandwidth and the relative gain classification of the application.

4.2 Transimpedance Amplifier Noise

While simple in structure, the transimpedance amplifier can exhibit a complex noise behavior. Basic noise components result from the circuit's feedback resistor and the amplifier input's noise current and noise voltage. The resistor and current noise sources produce immediately identifiable output noise components. However, the amplifier's input noise voltage often receives an unexpected high-frequency gain, resulting from a combination of high feedback resistance and high capacitance at the amplifier input. Parasitic capacitance around the feedback resistor and the characteristic $1/f$ response of the noise voltage further complicate the circuit's final noise response. Multiple poles and zeros shape this response, posing a formidable noise analysis task. However, breaking the analysis into a series of frequency regions, separated by the poles and zeros, restores simplicity. This regional analysis also permits the comparison of relative noise effects to identify circuit changes that would optimize performance. Following the individual analyses of these resistor, noise current, and noise voltage effects, a summarized listing of noise equations permits step-by-step analysis of specific applications.

4.2.1 Transimpedance amplifier noise sources[6]

Transimpedance amplifier output noise results from the noise effects of the circuit's feedback resistor and the op amp's input noise current and noise voltage. The resistor and current noise sources produce output noise components that follow from intuition. However, the amplifier's input noise voltage may receive an unexpected high-frequency gain due to the $1/\beta$ rise of Fig. 4.5. This particularly occurs where a large R_f feedback resistance or a large C_S source capacitance moves the f_z zero down into the frequency range of normal operation. As developed elsewhere,[7] an op amp circuit amplifies this noise voltage by a gain of $1/\beta$ up to the open-loop response roll off of the op amp.

For each of the circuit's noise sources, the complete noise evaluation begins by identifying the corresponding spectral noise densities and noise gains. Then rms integration incorporates frequency response effects for each of the three. As a noise measure, spectral noise density reflects the noise contained in a 1-Hz bandwidth at a given frequency and this density may vary with frequency. Rms integration sums the noise density effects over frequency, reflecting the broad-band noise of a given noise source. For the transimpedance amplifier, the spectral noise densities e_{noR}, e_{noi}, and e_{noe} represent the output noise components produced by the resistor itself, the amplifier's input current noise, and the amplifier's input voltage noise. These noise sources combine to produce the net output noise density e_{no}. As will be described, the e_{noR} and e_{noi} noise densities remain constant with frequency until a bandwidth limit truncates their effects. However, the e_{noe} noise density varies with frequency and rolls off at a different bandwidth limit.

Figure 4.7 models the basic transimpedance amplifier of Fig. 4.5 for noise analysis. In the model, capacitance C_{ia} and noise sources i_{ni} and e_{ni} represent the relevant characteristics of the amplifier input and

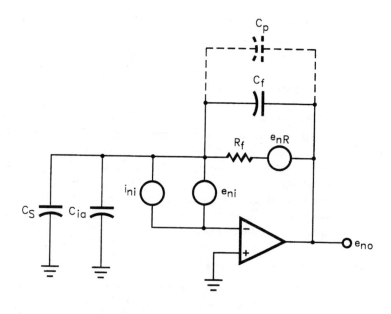

$$e_{no} = \sqrt{(e_{noR})^2 + (e_{noi})^2 + (e_{noe})^2}$$

$$e_{noR} = \sqrt{4KTR_f} \qquad e_{noi} = R_f\sqrt{2qI_{B-}} \qquad e_{noe} = \frac{1 + R_f(C_f + C_p + C_l)s}{1 + R_f(C_f + C_p)s}\, e_{ni}$$

$$C_i = C_S + C_{ia}, \quad C_{ia} = C_{id} + C_{icm}$$

Figure 4.7 Modeling Fig. 4.5 for noise analysis combines the amplifier noise sources e_{ni} and i_{ni} with the e_{nR} of the feedback resistor and the circuit capacitances that shape the noise response.

noise source e_{nR} represents the noise voltage of the feedback resistor. The noise contributed directly by the feedback resistor displays a spectral density of[8] $e_{nR} = \sqrt{4KTR_f}$, where K is Boltzmann's constant $(1.38 \times 10^{-23}$ J/K) and T is temperature in kelvin (°C + 273). This resistor noise voltage transfers to the output of a current-to-voltage converter without amplification, making $e_{noR} = e_{nR}$. Indirectly, the R_f feedback resistor also influences the transimpedance amplifier's noise through interaction with the op amp's noise sources. Noise current i_{ni} represents the shot noise of the input bias current I_{B-} and displays a noise density of[9] $i_{ni} = \sqrt{2qI_{B-}}$, where q is the charge on an electron $(1.6 \times 10^{-19}$ coulomb). This noise current flows directly through the feedback resistor, producing a noise voltage of $e_{nRi} = i_{ni}R_f = R_f\sqrt{2qI_{B-}}$. Like the noise voltage of R_f itself, this current noise effect transfers to the circuit output with unity gain and $e_{noi} = e_{nRi}$.

In a more complex relationship, the amplifier's input noise voltage e_{ni} often receives an amplification characterized by a noise gain peaking. At first, it might seem that e_{ni} would also transfer to the circuit output with low gain, and indeed at dc, the circuit presents a noise voltage gain of $A_{ne} = 1$ to e_{ni}. However, at higher frequencies, capacitances modify this noise gain. In the figure C_S, C_{ia}, C_f, and C_p represent the capacitances of the source, the amplifier input, the phase compensation, and parasitic feedback shunting, respectively. At higher frequencies, these capacitances convert the noise gain received by e_{ni} to

$$A_{ne} = \frac{1}{\beta} = \frac{1 + R_f(C_f + C_p + C_I)s}{1 + R_f(C_f + C_p)s}$$

This noise gain continues up to the boundary imposed by the op amp's open-loop response. At that point, A_{ne} rolls off with that response becoming $A_{ne} = A_{OL} = f/f_c$, where f_c is the unity-gain crossover frequency of the op amp. For each of the three A_{ne} noise gains described here, the output noise resulting from e_{ni} becomes $e_{noe} = A_{ne}e_{ni}$.

4.2.2 Transimpedance amplifier total noise

The spectral nature of the preceding noise densities restricts numerical noise analysis to one frequency at a time. Broad-band evaluation of output noise requires summing the individual frequency effects over the bandwidth of the amplifier. Simple rms integration performs this summation, yielding noise analysis equations for the transimpedance amplifier. Gain and bandwidth affect the three fundamental noise sources differently and first require separate rms analyses of individual output noise effects. Then, an rms combination of the three separate effects defines the total output noise. In this process, the separate analyses permit identification of the predominant noise effect in a given application. The discussion that follows first considers the simpler effects of e_{noR} and e_{noi}. A later examination defines the differences that extend this process to the third noise component, e_{noe}.

Rms analysis summarizes the effect of a given noise source by mathematically integrating its spectral noise effect over frequency. This integration combines the circuit's noise density, gain, and bandwidth characteristics into a single noise indicator, the rms value of the total noise. For each noise source, converting noise density to rms noise requires evaluating the integral[2]:

$$E_{no}^2 = \int_0^\infty |A_n e_n|^2 \, df$$

Here, e_n represents an input-referred noise source and A_n represents the corresponding noise gain supplied to e_n. Noise gain A_n includes the relevant bandwidth characteristic.

The constant noise gains and constant noise densities associated with the circuit's feedback resistor and the amplifier's input current simplify the integration for these two noise sources. There, constant noise gains of $A_n = 1$ ideally transfer the effects of these two noise sources directly to the circuit output, developing the output noise densities e_{noR} and e_{noi} described before. However, the bandwidth limit of the circuit's transresistance rolls off this ideal transfer response at some higher frequency. For high-gain current-to-voltage converters, this bandwidth limit occurs where the parasitic capacitance C_p breaks with R_f. In other cases, the addition of a C_f phase compensation capacitance in parallel with C_p modifies this break frequency. In either case, this break frequency simultaneously rolls off the output signals developed by the input signal, the resistor noise, and the amplifier's current noise. Then, $A_n = 1/(1 + jf/\text{BW})$, where $\text{BW} = 1/2\pi R_f (C_p + C_f)$, and this defines one variable of the preceding integral. The other variable, e_n, follows from the analyses of Sec. 4.2.1 for the resistor and current noise effects. There, $e_{nR} = \sqrt{4KTR_f}$ and $e_{nRi} = R_f\sqrt{2qI_{B^-}}$ describe the constant noise densities produced by the feedback resistor and amplifier input noise current. Evaluating the rms integral for these A_n and e_n conditions defines the rms output noise contributions as

$$E_{noR} = \sqrt{2KTR_f\pi\text{BW}}$$

and

$$E_{noi} = R_f\sqrt{q\pi\text{BWI}_{B^-}}$$

where $\text{BW} = 1/2\pi R_f (C_p + C_f)$.

Constant noise densities, constant noise gains, and a common noise bandwidth simplified the rms integration of the e_{nR} and e_{nRi} noise voltages. The e_{ni} noise voltage, however, displays a variable noise density, experiences a frequency-dependent amplification, and receives a greater noise bandwidth than e_{nR} and e_{nRi}. Amplifier noise voltage e_{ni} exhibits a characteristic $1/f$ response at lower frequencies, and the corresponding noise gain A_{ne} of the circuit increases at higher frequencies. Also, the capacitance bypass of R_f, which set the bandwidth BW in the foregoing, levels but does not roll off the A_{ne} noise gain received by e_{ni}. Only the open-loop response of the op amp finally rolls off this noise gain. As a result, the $e_{noe} = A_{ne}e_{ni}$ product of the rms integral becomes a complicated function of frequency.

Figure 4.8 summarizes the frequency response of $e_{noe} = A_{ne}e_{ni}$ for the general case where $f_p < f_i$. At low frequencies, this response begins

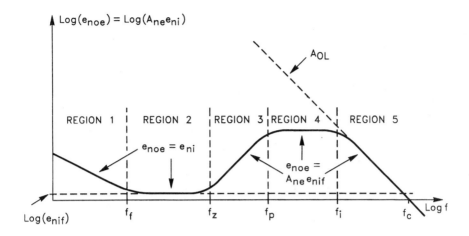

$$f_z = 1/2\pi R_f(C_f + C_p + C_I) \quad f_p = 1/2\pi R_f(C_f + C_p) \quad f_i = f_c(C_f + C_p)/(C_f + C_p + C_I)$$

$$E_{noe} = \sqrt{E_{noe1}^2 + E_{noe2}^2 + E_{noe3}^2 + E_{noe4}^2 + E_{noe5}^2}$$

Figure 4.8 Output noise e_{noe}, resulting from e_{ni}, displays a complex frequency response that encourages breaking the response into linear regions for rms analysis of E_{noe}.

with a declining magnitude resulting from the $1/f$ noise of e_{ni}. In this initial region, the noise power declines in proportion to $1/f$ and, thus, the noise voltage e_{ni} declines in proportion to $\sqrt{1/f}$. In this region,[10]

$$e_{ni} = e_{nif}\sqrt[4]{1 + (f_f/f)^2}$$

The $1/f$ decline continues up to the noise-floor corner at f_f, after which e_{ni} assume the noise floor level as $e_{ni} = e_{nif}$. At these lower frequencies, the transconductance amplifier supplies the unity noise gain expected from the circuit's single resistor feedback and the corresponding noise gain of $A_{ne} = 1/\beta = 1$. Following f_f, this unity gain continues temporarily with the e_{noe} response following the noise floor level at e_{nif}. However, A_{ne} later increases beginning at the $f_z = 1/2\pi R_f(C_f + C_p + C_I)$ zero of the $1/\beta$ response, as derived before. This zero makes $A_{ne} \approx f/f_z$ up to the $1/\beta$ pole at $f_p = 1/2\pi R_f(C_f + C_p)$, as also derived before. Following this pole, the circuit's capacitances temporarily control the feedback, producing a level noise gain of $A_{ne} = 1/\beta = 1 + C_I/(C_f + C_p)$. This level $1/\beta$ response intercepts $A_{OL} \approx f/f_c$ at $f_i = f_c(C_f + C_p)/(C_f + C_p + C_I)$. Following this intercept, $A_{ne} \approx A_{OL} \approx f/f_c$.

The varying e_{ni} and A_{ne} responses described in the preceding would permit expressing e_{noe} as a single equation. However, just examining the associated curve of Fig. 4.8 indicates the very complex result that the corresponding rms integration would produce. Instead, separating the integration into the five response regions illustrated reduces the integration task to a more manageable and more insightful evaluation.[11] Then, approximation segments permit manual analysis and reveal the relative noise effects of different frequency ranges. This process produces the rms results E_{noe1} through E_{noe5}, which represent the noise effect of e_{ni} in the five response regions for later rms combination with the e_{noR} and E_{noi} results derived before. The summary that follows defines the E_{noe} results of the approximation process.

4.2.3 Transimpedance amplifier noise summary

Rms summations combine the seven noise components of the preceding sections to define the transimpedance amplifier's net output noise E_{no}. First, combining the five components of E_{noe} in this manner defines the net rms noise produced by op amp input noise voltage at the transimpedance amplifier output in E_{noe}. Then, combining the E_{noe} result with the e_{noR} and E_{noi} results of Sec. 4.2.2, produces the total output noise E_{no} of the transimpedance amplifier. Once again, rms summation combines the noise components. To simplify computation of a specific E_{no} result, the listing that follows summarizes the seven fundamental noise components described before. These equations depend on other expressions that define noise magnitudes, cutoff frequencies, and capacitances. For simplicity, the listing defines these other expressions only at their first occurrence. Step-by-step calculation of E_{no} for the basic transimpedance amplifier proceeds through the individual calculations:

$E_{noe1} = e_{nif}\sqrt{f_f \ln(f_f/f_1)}$, where e_{nif} is the e_{ni} noise floor level and f_f is the $1/f$ noise corner frequency. Here, setting $f_1 \approx 0.01$ Hz instead of 0 Hz avoids a theoretical contradiction of $1/f$ noise theory[12,13] and still ensures good analytical accuracy.

$E_{noe2} = e_{nif}\sqrt{f_z - f_f}$, where $f_z = 1/2\pi R_f(C_f + C_p + C_I)$, and $C_I = C_S + C_{id} + C_{icm}$ and C_p is the parasitic feedback capacitance. Here and in the E_{noe4} analysis that follows, numerical evaluation can produce the square root of a negative number. Then substitute zero for the result as explained after this summary.

$E_{noe3} = (e_{nif}/f_z)\sqrt{(f_p^3 - f_z^3)/3}$, where $f_p = 1/2\pi R_f(C_f + C_p)$.

$E_{noe4} = [1 + C_I/(C_f + C_p)]\, e_{nif}\sqrt{f_i - f_p}$, where $f_i = f_c(C_f + C_p)/(C_f + C_p + C_I)$.

$E_{noe5} = (e_{nif}/f_c)\sqrt{1/f_i}$, where f_c is the op amp's unity-gain crossover frequency.

$E_{noe} = \sqrt{E_{noe1}^2 + E_{noe2}^2 + E_{noe3}^2 + E_{noe4}^2 + E_{noe5}^2}$, the total rms output noise resulting from the noise voltage density e_{ni} of the op amp input.

$E_{noR} = \sqrt{2KTR_f\pi BW}$, where $BW = 1/2\pi R_f(C_f + C_p)$.

$E_{noi} = R_f \sqrt{q\pi BW\, I_{B-}}$.

$E_{no} = \sqrt{E_{noR}^2 + E_{noi}^2 + E_{noe}^2}$.

Special cases potentially affect the E_{noe2} and E_{noe4} results. These noise components cover flat response regions in Fig. 4.8 that certain circuit combinations eliminate. In that figure, the condition $f_z < f_f$ eliminates region 2 and $f_i < f_p$ eliminates region 4. Very large R_fC_p time constants can produce one or both of these conditions. Then, the eliminated region or regions produce zero noise contribution. However, to cover all possible cases, the segmented E_{noe} analysis must include the potential region 2 and region 4 noise contributions. Numerical evaluations of the E_{noe2} and E_{noe4} equations easily identify these special cases by delivering the square root of a negative number. Then simply substitute zero for the region's numerical result. When this results for E_{noe4}, it signals the potential need for the phase compensation described in Sec. 4.1.3.

For a given application, the numerical evaluation of E_{no}, following the foregoing listing, identifies the relative significance of the individual noise contributors. There, numeric comparison of terms in both the E_{noe} and the E_{no} calculations reveals the actual factors that affect noise performance significantly. Without these comparisons, traditional intuition alone leads to suboptimal design choices. For example, intuition biases the op amp selection toward low input noise voltage. However, this selection bias risks unnecessarily sacrificing other op amp characteristics. By itself, the amplifier's noise voltage only dominates noise performance in a portion of transimpedance amplifier designs. In other cases, noise from the feedback resistor dominates, making the amplifier noise secondary. In still other cases, the noise combination of the amplifier noise and the feedback resistance dominates noise performance through noise gain peaking. The multidimensional aspects of this noise performance precludes fixed guidelines that would replace numerical evaluation and comparison.

4.3 Basic Transconductance Amplifiers

As the name suggests, transconductance amplifiers convert an input signal voltage into an output current. Common implementations of this function derive from the difference amplifier circuit with modifications that introduce positive feedback to produce the voltage-to-current conversion. In the simplest case, the difference amplifier connected for

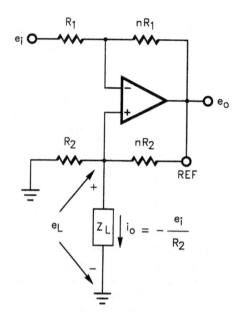

Figure 4.9 Difference amplifier with positive feedback performs a transconductance role to deliver an output current in response to an input voltage.

both positive and negative feedback supplies an output current proportional to an input voltage. The two feedback connections of this circuit would complicate performance analysis but employing feedback analysis easily defines the transfer response, output offset, bandwidth, and terminal impedances of the basic transconductance amplifier.

4.3.1 Transconductance amplifier transfer response

In its simplest form, the difference amplifier of Sec. 1.3 serves the transconductance role with just three modifications, as shown in Fig. 4.9. There, the difference amplifier's reference pin first connects to the amplifier output to establish a positive feedback that removes the circuit's sensitivity to load voltage. Next, a load Z_L connects to the amplifier's noninverting input to receive the feedback-corrected output current of the circuit. Finally, different rather than equal resistor values define the circuit's R_1 and R_2 resistances. As will be seen, this resistance difference permits independent control of transconductance and input impedance.

Normally, the Z_L connection shown would disturb the precise impedance balance required for the difference amplifier. However, in this case, the positive feedback corrects for the effect of the imbalance. Voltage developed upon load Z_L acts as a signal at the op amp's noninverting input, and circuit amplification of this signal adjusts the op amp output voltage to accommodate the load voltage effect. The added output voltage supplies a correction current through the positive feedback network formed by resistors nR_2 and R_2 in conjunction with the load Z_L. Matching the circuit's two $n{:}1$ resistor ratios makes the correction current supplied to Z_L accurately compensate for the effect of the load, even though circuit impedances remain imbalanced.

Beyond this qualitative evaluation, the dual feedback connections of this circuit obscure any intuitive anticipation of its quantitative performance. The circuit structure offers little insight into the transfer response and bandwidth. Further, the circuit's positive feedback raises a question of frequency stability, and the dynamic range restriction introduced by the amplifier output swing limitation poses another question. Straightforward analysis of all of these performance characteristics represents a formidable task. However, feedback modeling reduces this task to voltage divider analysis through the information derived from the feedback and feedforward factors of the circuit. This modeling reduces the transfer response of any op amp circuit to the standard equation[14]

$$A_{CL} = \frac{e_o}{e_i} = \frac{-\alpha/\beta}{1 + 1/A\beta} = \frac{A_{CLi}}{1 + 1/A\beta}$$

In this equation, A represents the op amp's open-loop gain, β the net feedback factor, and α the input feedforward factor. Also, $A_{CLi} = -\alpha/\beta$ defines the ideal closed-loop voltage gain of the e_o/e_i response. At lower frequencies, the $A\beta$ loop gain of the equation remains high, reducing the denominator to unity for $A_{CL} \approx A_{CLi}$.

This A_{CL} response defines the op amp output voltage e_o rather than the desired output signal i_o. However, a straightforward feedback analysis yields the desired i_o result. That analysis begins with the ideal result $A_{CLi} = e_o/e_i = -\alpha/\beta$, which first yields $e_o = -(\alpha/\beta)e_i$. Then, the circuit's positive feedback converts the e_o result to the load voltage e_L. That factor, β_+, equals the fraction of the op amp output voltage fed back to the amplifier's noninverting input, making $e_L = (\beta_+)e_o = -(\alpha\beta_+/\beta)e_i$. Next, substituting the relationship $e_L = i_o Z_L$ defines the output load current in terms of feedback quantities as

$$i_o = -\frac{\alpha\beta_+}{\beta Z_L}e_i$$

This expression reduces the circuit's response to the desired relationship between i_o and e_i. However, to be useful, it requires the definition of α, β_+, and β in terms of the circuit's feedback elements, as determined next.

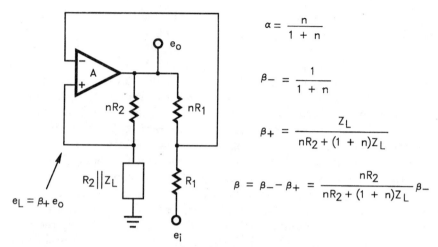

Figure 4.10 Redrawing the Fig. 4.9 transconductance amplifier displays the circuit's feedback voltage dividers for feedback analysis.

4.3.2 Transconductance amplifier feedback analysis

Continued feedback analysis defines the final i_o expression through simple divider ratios. Redrawing the transconductance amplifier in Fig. 4.10 begins the analysis process that defines the circuit's α and β factors. In the figure, drawing the circuit's feedback networks as simple voltage dividers adds insight into the analysis. These dividers supply the op amp inputs with attenuated signals from the amplifier output and from the signal input. As shown, combining R_2 and Z_L reduces the positive feedback network to a simple divider with one divider element equaling $R_2 \| Z_L$. Then, simple two-element divider ratios define the α and β factors for the circuit. The α feedforward factor equals the voltage divider ratio of the negative feedback network as presented to the e_i signal input. Assuming the zero output impedance of the ideal op amp output, this divider ratio becomes

$$\alpha = \frac{n}{1 + n}$$

Next, the β feedback factor equals the fraction of the output fed back to the op amp input. For most op amp circuits a single feedback network defines this factor through its negative feedback connection between the output and the op amp's inverting input. Then, the voltage divider ratio seen from output to input defines the feedback factor. This transconductance amplifier retains that connection to produce a negative feedback factor of

$$\beta_- = \frac{1}{1 + n}$$

However, this amplifier also couples positive feedback to the op amp's noninverting input, modifying the circuit's net feedback factor. The latter feedback couples through the network formed by nR_2 and $R_2 \| Z_L$, and this produces the positive feedback factor

$$\beta_+ = \frac{Z_L}{nR_2 + (1 + n)Z_L}$$

The differential inputs of the op amp respond to the difference between the signals fed back to the inverting and noninverting inputs, making the circuit's net feedback factor $\beta = \beta_- - \beta_+$. Subtracting the preceding β_- and β_+ results defines the net circuit feedback factor as

$$\beta = \frac{nR_2}{nR_2 + (1 + n)Z_L} \beta_-$$

Substituting these various factors into the $i_o = -(\alpha\beta_+)e_i/\beta Z_L$ result of Sec. 4.3.1 expresses that result in terms of the circuit's feedback elements as simply

$$i_o = -\frac{e_i}{R_2}$$

Thus only the R_2 resistance value defines the ideal transconductance of the circuit. As expressed, this ideal transconductance remains independent of the load impedance Z_L, reflecting an infinite impedance at the circuit's current output terminal. Also, this expression shows no dependence on the R_1 resistance level. As will be seen, the latter fact permits an independent setting of the circuit's input resistance.

4.3.3   Transconductance amplifier performance analysis

The feedback analysis of Sec. 4.3.2 defines feedback equations that also determine the offset, bandwidth, stability, and signal swing limitations of the basic transconductance amplifier. The circuit's transfer response guides the determination of output offset current, the circuit's net β determines bandwidth and stability, and the β_+ equation defines the maximum swing limitation of i_o.

First, consider the output offset current as produced by the op amp's input offset voltage and input bias currents, modeled in Fig. 4.11. Superposition guides the associated analysis considering one effect at a time, first by grounding the signal input as shown. Then,

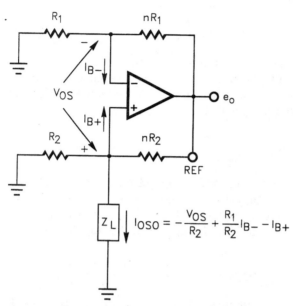

Figure 4.11 Op amp dc input errors produce an output off-set current for the basic transconductance amplifier.

considering the effect of the op amp's input offset voltage V_{OS} in the absence of the I_B currents defines a first component of the output offset current. Under these conditions, V_{OS} controls the voltage of the input circuit and excites the circuit's normal transconductance gain of $i_o = -e_i/R_2$ to produce an output offset component equal to $-V_{OS}/R_2$. Next, consider the effect of I_{B-} in the absence of V_{OS} and I_{B+}. With $V_{OS} = 0$, no voltage exists in the input circuit and feedback forces the nR_1 resistor to accept the I_{B-} current. In this process, feedback replicates the resulting voltage on the nR_2 resistance, developing a current equal to $(R_1/R_2)I_{B-}$. This current flows directly to the Z_L load because the $V_{OS} = 0$ condition prevents current flow in the R_2 resistance of the input circuit. Finally, consider the effect of I_{B+} in the absence of V_{OS} and I_{B-}. Under these conditions, the $V_{OS} = 0$ condition and the $I_{B-} = 0$ condition combine to prevent this current from flowing in either R_2 or nR_2 and I_{B+} flows directly to the Z_L load. Together, the three op amp offset errors produce an output offset current of

$$I_{OSO} = -\frac{V_{OS}}{R_2} + \frac{R_1}{R_2}I_{B-} - I_{B+}$$

Also from Sec. 4.3.2, the circuit's net feedback factor predicts bandwidth for resistively loaded cases. In an earlier publication,[15] analysis defined the bandwidth of resistive feedback op amp circuits as βf_c, where f_c is the frequency of the amplifier's unity-gain crossover. The potentially re-active Z_L of this present case precludes general use of this simplification but, for the common case of a resistive load, $Z_L = R_L$ and the circuit's net feedback factor remains controlled by resistive elements. Translating the previous β result for the $Z_L = R_L$ condition produces

$$\beta = \frac{nR_2}{nR_2 + (1 + n)R_L}\beta_-$$

where $\beta_- = 1/(1 + n)$. Under this condition, the circuit produces the traditional bandwidth BW $= \beta f_c$. The positive feedback introduced here reduces the bandwidth by reducing the net circuit β. From the β equation, the positive feedback makes the net feedback factor a fraction of the negative feedback factor β_-. This fraction depends on the load resistance R_L and approaches zero as R_L becomes large, making the bandwidth approach zero.

The initial β equation of Sec. 4.3.2 also reveals stability information, both by itself and in conjunction with traditional graphical analysis. By itself, the equation

$$\beta = \frac{nR_2}{nR_2 + (1 + n)Z_L}\beta_-$$

where $\beta_- = 1/(1 + n)$ resolves the question of oscillation resulting from the positive feedback. Examination of this equation shows that β retains a positive polarity for all positive values of Z_L, and this assures that negative feedback prevails for any practical load impedance. Otherwise, the positive feedback would dominate the circuit to induce oscillation or latching. However, even with a dominant negative feedback, the potentially reactive nature of Z_L can still induce oscillation. Further stability information results from graphical analysis using the β result in feedback stability analysis. As described previously,[16] plotting the $1/\beta$ curve with the amplifier's open-loop gain response permits an evaluation of feedback stability, and this analysis includes the frequency dependence of β.

Finally, the β_+ expression derived in Sec. 4.3.2 relates the signal swing maximum of i_o to that of the op amp output voltage e_o. As described in Sec. 4.3.1, the load voltage e_L equals the fraction of the output voltage fed back to the amplifier's noninverting input. This fraction represents the circuit's positive feedback and $e_L = \beta_+ e_o$. Combining this with $e_L = i_o Z_L$ and solving yields $i_o = \beta_+ e_o / Z_L$, which reflects the dependence of i_o on e_o. Output voltage e_o encounters a maximum swing limitation at some $e_{o\,max}$, where the amplifier output stage reaches its saturation limit. This limit simultaneously imposes an output current limit for the basic transconductance amplifier. Substituting the previously derived expression for β_+ defines this limit in terms of circuit elements as

$$i_{o\,max} = \frac{e_{o\,max}}{nR_2 + (1 + n)Z_L}$$

4.3.4 Transconductance amplifier terminal impedances

Ideally, a transconductance amplifier would present infinite impedances at both its input and its output terminals. An infinite input impedance prevents loading errors on the input voltage source and an infinite output impedance prevents output current change due to output voltage swing. The basic transconductance amplifier approximates this ideal with finite impedances determined by circuit impedances and their imbalance. Figure 4.12 models this amplifier for analyses of these impedances with the i_i input current that limits input impedance and a δn resistance imbalance that limits output impedance.

First, consider the finite input impedance defined by the i_i input current drawn in response to an e_i input voltage. For this analysis the δn modification can be ignored, as this resistance error only affects output impedance. By definition, input impedance equals e_i divided

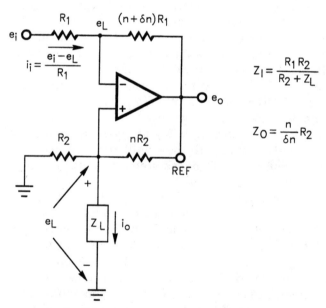

Figure 4.12 Feedback impedances and impedance imbalance δn define the input and output impedances of the basic transconductance amplifier.

by the resulting input current, or $Z_I = e_i / i_i$. Signal e_i initiates this current by driving one end of the R_1 resistor. However, this signal alone does not control the voltage across that resistor. The circuit's output current i_o develops the voltage e_L at the op amp's noninverting input, and feedback replicates that voltage at its inverting input, thus controlling the other end of the R_1 resistor. The resulting input current becomes $i_i = (e_i - e_L)/R_1$. From Sec. 4.3.2, $e_L = i_o Z_L = -(e_i/R_2)Z_L$, and combining these equations yields the input impedance

$$Z_I = \frac{e_i}{i_i} = \frac{R_1 R_2}{R_2 + Z_L}$$

Note that Z_I depends on both R_1 and R_2, and this provides an extra degree of freedom in the design control of input impedance. As demonstrated in Sec. 4.3.2, R_2 must be chosen to set the circuit's transconductance, reflected in the $i_o = -e_i/R_2$ transfer response. With the conventional difference amplifier, $R_1 = R_2$, and setting this circuit's $1/R_2$ transconductance would automatically define input impedance. However, making the R_1 and R_2 base resistance values different in this case establishes a second degree of freedom for independent control of the input impedance. Selecting R_2 still defines the circuit's transconductance, but selecting R_1 provides independent control of the input impedance. A practical factor sometimes limits this freedom through the shunting effects of parasitic capacitances. In such cases, very large values for R_1 or R_2 potentially jeopardize the impedance balance required to maintain high output impedance.

For the typical case, just resistance imbalance limits that output impedance through the δn ratio imbalance modeled in Fig. 4.12. In the ideal case, $\delta n = 0$ and the analysis of Sec. 4.3.2 produces $i_o = -e_i/R_2$. This ideal transfer response reflects no output current sensitivity to load impedance, indicating infinite output impedance. That condition requires perfect matching of the circuit's two n:1 resistance ratios, but in practice, resistance tolerance errors disturb these ratios. Tolerance errors in all four of the R_1 and R_2 resistances contribute to this imbalance, but combining their net effects in the δn shown simplifies the analysis and delivers an equivalent result. Then, δn represents the net imbalance of the circuit's two n:1 resistance ratios.

Analyzing the circuit's transfer response with δn included defines the output impedance of the practical circuit. Section 4.3.1 found this transfer response to be

$$i_o = -\frac{\alpha \beta_+}{\beta Z_L} e_i$$

Then, the analysis of Sec. 4.3.2 derived the α and β factors of this expression to yield $i_o = -e_i/R_2$. Repeating the latter analysis with δn in place but assuming $\delta n \ll n$ produces

$$i_o = -\frac{e_i}{R_2 - Z_L\,\delta n/n}$$

This actual transfer response displays a sensitivity to the load impedance Z_L, indicating a finite output impedance.

To quantify that impedance, a substitution, a derivative, and then an inversion define Z_O. First, rearranging $e_L = i_o Z_L$ produces $Z_L = e_L/i_o$ for substitution in the last expression. Then solving for i_o yields

$$i_o = \frac{(\delta n)e_L - ne_i}{nR_2}$$

Next, differentiating this expression defines the output current sensitivity to load voltage e_L as $di_o/de_L = \delta n/nR_2$. However, di_o/de_L represents a conductance rather than an impedance. Inverting this expression produces de_L/di_o, or the circuit's output impedance of

$$Z_O = \frac{n}{\delta n} R_2$$

Thus the same R_2 resistance that defines the circuit's transconductance through $i_o = -e_i/R_2$ also defines output impedance. For a given $\delta n/n$, decreasing R_2 to increase transconductance also decreases Z_O in direct proportion.

4.4   Improved Transconductance Amplifiers

As described in the preceding section, the basic transconductance amplifier requires precise resistance matching to achieve high output impedance. This normally requires resistance trimming of individual circuits, but pretrimmed difference amplifiers automatically provide the required matching. However, these pretrimmed amplifiers do not normally provide the access to the op amp's noninverting input, as required for the connection of the basic transconductance amplifier's load. To accommodate these amplifiers, an improved transconductance amplifier moves the load to the circuit output through the addition of a bootstrap voltage follower. This improvement retains the independent control of input impedance and transconductance and greatly simplifies circuit analysis. Still, this alternative includes two op amps in a common feedback loop, requiring greater attention to bandwidth and stability constraints.

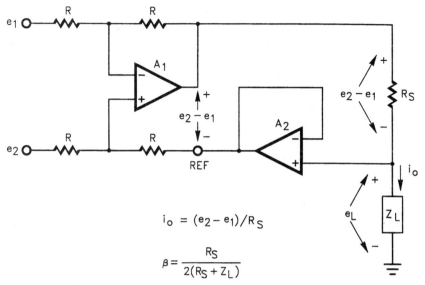

$$i_o = (e_2 - e_1)/R_S$$

$$\beta = \frac{R_S}{2(R_S + Z_L)}$$

Figure 4.13 Adding an R_S sense resistor and a bootstrap voltage follower to a pretrimmed difference amplifier forms a transconductance amplifier without the need for individual circuit adjustment.

4.4.1 Adding a bootstrap follower

Figure 4.13 shows the improved connection as a difference amplifier formed with A_1 accompanied by a bootstrap voltage follower formed with A_2. Most pretrimmed difference amplifiers present unity voltage gain and the example here illustrates this case. From the discussion of Sec. 1.3, the unity-gain difference amplifier normally supplies an output voltage equal to $e_2 - e_1$. That condition continues here with the A_1 op amp producing $e_2 - e_1$ between its output and the reference terminal of its associated difference amplifier. Voltage follower A_2 drives this reference terminal in a positive feedback connection that replicates that output voltage across the sense resistor R_S. The resulting current in R_S cannot flow into the A_2 input, and the circuit's feedback forces this current into the load Z_L. This bootstrap condition produces the voltage-to-current conversion for an output current of

$$i_o = \frac{e_2 - e_1}{R_S}$$

The impedance isolation of the voltage follower simplifies performance analyses in the determination of bandwidth, offset, and terminal impedances. As will be seen, the selection of the circuit's R_S sense resistor directly influences all of these performance characteristics.

Through the isolation of the voltage follower, the circuit presents only two-element voltage dividers for analysis. This simplification first aids the determination of bandwidth through feedback factor analysis. As before, this circuit employs both positive and negative feedback, and the combination determines the net feedback factor. The negative feedback to the inverting input of A_1 produces $\beta_- = 0.5$. The positive feedback to the noninverting input of A_1 first couples through R_S and Z_L encountering an attenuation equal to $Z_L/(Z_L + R_S)$. Next, this attenuated signal couples through the voltage follower and, assuming $f_{c2} \gg f_{c1}$, this coupling occurs with unity gain throughout the useful frequency range of the circuit. The voltage follower output drives the equal-resistance voltage divider connected to the noninverting input of A_1 through which the signal encounters an additional attenuation factor of 0.5 for a net positive feedback factor of

$$\beta_+ = \frac{Z_L}{2(R_S + Z_L)}$$

Together, the negative and positive factors produce a net feedback factor of $\beta = \beta_- - \beta_+$, or

$$\beta = \frac{R_S}{2(R_S + Z_L)}$$

For most op amp circuits, the net feedback factor defines the circuit's signal bandwidth through the simple relationship $BW = \beta f_c$, where f_c is the unity-gain crossover frequency of the op amp. However, the Z_L of the preceding expression potentially makes β a function of frequency, often requiring a more detailed feedback analysis.[17] Still for resistively loaded cases, $Z_L = R_L$, and the simplified relationship defines the bandwidth as

$$BW = \frac{f_c}{2(1 + R_L/R_S)}$$

From this expression, note that larger R_L load resistances decrease bandwidth while larger R_S sense resistances increase it. This produces a compromise with output voltage swing requirements. The voltages developed upon both R_L and R_S depend on the inherently limited output voltage swing of A_1. Making $R_L \gg R_S$ reserves most of this voltage swing for the final circuit output. However, this condition reduces bandwidth by reducing the feedback factor. As will be seen later, output offset and terminal impedance requirements place further demands upon R_S, making the selection of this resistance a multidimensional compromise.

Two factors potentially alter the preceding bandwidth result arising from the assumptions made in the foregoing analysis. There, the assumptions $f_{c2} \gg f_{c1}$ and $Z_L = R_L$ serve to demonstrate the basic bandwidth performance. Then the $f_{c2} \gg f_{c1}$ assumption assures that the voltage follower supplies a simple gain of $A_{CL2} = 1$ over the useful frequency range of the circuit and the $Z_L = R_L$ assumption permits use of the BW $= \beta f_c$ simplification. These assumptions hold for many implementations of the improved transconductance amplifier. However, other application conditions make $f_{c2} \approx f_{c1}$ or introduce a reactive Z_L, requiring more detailed feedback analysis[18] to determine the circuit's bandwidth.

The impedance isolation of the voltage follower also simplifies offset analysis, but this added amplifier introduces new sources to the output offset current. Figure 4.14 models the improved transconductance amplifier for offset analysis showing the individual errors of the two op amps and their combined effect in I_{OSO}. There, superposition grounding of the signal inputs focuses the analysis upon the offset effects. As shown, the V_{OSO1} output offset of the difference amplifier combines with the V_{OS2} input offset of the voltage follower to produce the voltage $V_S = V_{OSO1} - V_{OS2}$ on the R_S sense resistor. In this expression, V_{OSO1} represents the output offset voltage of the difference amplifier, as developed by the input offset errors of A_1. Those errors produce $V_{OSO1} = -2V_{OS1} + (I_{B1-} - I_{B1+})R$. The resulting current in R_S combines with the I_{B2+} input bias current of A_2 to produce the output offset current

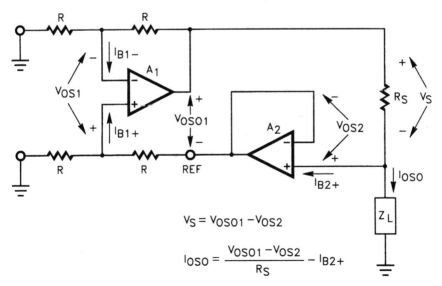

$$V_S = V_{OSO1} - V_{OS2}$$

$$I_{OSO} = \frac{V_{OSO1} - V_{OS2}}{R_S} - I_{B2+}$$

Figure 4.14 The improved transconductance amplifier of Fig. 4.13 produces an output offset current determined by the input offset errors of two op amps.

$$I_{OSO} = \frac{V_{OSO1} - V_{OS2}}{R_S} - I_{B2^+}$$

Note that making R_S small, to develop high transconductance, makes I_{OSO} larger, paralleling the familiar gain-versus-offset compromise of typical op amp circuits.

4.4.2 Bootstrapped transconductance amplifier terminal impedances

Three impedances characterize the two inputs and the output of the bootstrapped transconductance amplifier. Each of these depends on the circuit's positive feedback through the feedback signal coupled to the A_1 noninverting input. Figure 4.15 models the amplifier for input impedance analysis, and there the voltage follower again reproduces the load voltage e_L at the reference pin of the difference amplifier. This makes the lower two R resistors a voltage divider driven from both ends. Together e_2 and e_L drive this divider to produce the signal $e'_{icm} = (e_2 + e_L)/2$ at the noninverting input of A_1. Feedback replicates this signal at the inverting input of A_1, making it a common-mode input signal for this op amp. Here the notation e'_{icm} represents the common-mode signal reaching the inputs of the difference amplifier's op amp, as described in Sec. 1.3.2.

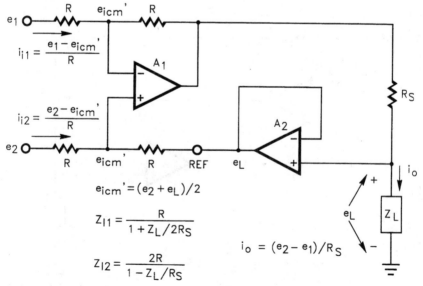

Figure 4.15 Input currents, drawn in response to the circuit's two input signals, define the input impedances of the improved transconductance amplifier of Fig. 4.13.

The presence of e'_{icm} alters the i_{i1} and i_{i2} input currents drawn at the e_1 and e_2 inputs and, thus, alters the associated input impedances. In this way, the e_L influence on e'_{icm} affects these impedances. Analyzing the circuit for $Z_{I1} = de_1/di_{i1}$ produces

$$Z_{I1} = \frac{R}{1 + Z_L/2R_S}$$

Note from this expression that as Z_L approaches zero, removing the positive feedback effect, Z_{I1} reverts to the $Z_{I1} = R$ of the basic difference amplifier. Analyzing the circuit for $Z_{I2} = de_2/di_{i2}$ produces

$$Z_{I2} = \frac{2R}{1 - Z_L/R_S}$$

Again as Z_L approaches zero, Z_{I2} reverts to the $Z_{I2} = 2R$ of the basic unity-gain difference amplifier.

Note that R_S affects both input impedances directly. Previously, the transfer response $i_o = (e_2 - e_1)/R_S$ expressed the circuit's transconductance as $1/R_S$ and this precludes selecting R_S for input impedance control. However, this dual influence does not remove the design independence of transconductance and input impedance. The difference amplifier's R resistance level remains an independent variable that also affects Z_{I1} and Z_{I2}.

The circuit's positive feedback establishes a high output impedance limited by the finite CMRR of the difference amplifier. Figure 4.16 models the circuit for analysis of this impedance. There, superposition grounding of the signal inputs permits analysis focus upon the influence of the load voltage on output current. In the figure, output current i_o develops the load voltage e_L, and the voltage follower transfers this voltage to the lower feedback network of the difference amplifier. Under the conditions shown, this voltage develops $e'_{icm} = e_L/2$. This e'_{icm} notation permits relating the circuit here with the basic difference amplifier described in Sec. 1.3.2. There, analysis defined $e'_{icm} = e_{icm}/2$ in response to an e_{icm} input signal. For the unity-gain case considered here, that analysis also defines a resulting output error of $e_{icm}/\mathrm{CMRR}_{DA} = 2e'_{icm}/\mathrm{CMRR}_{DA}$, where CMRR_{DA} is the common-mode rejection of the difference amplifier. In an equivalent condition, the common-mode signal produced by the positive feedback here produces $e'_{icm} = e_L/2$ for a difference amplifier output error of e_L/CMRR_{DA}. Feedback through the voltage follower reproduces this signal across R_S to develop an output error current of $i_o = e_L/\mathrm{CMRR}_{DA}R_S$. Solving this result for $Z_O = de_L/di_o$ produces the output impedance result

$$Z_O = \mathrm{CMRR}_{DA}R_S$$

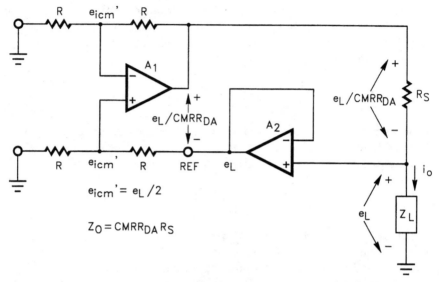

Figure 4.16 Superposition analysis grounds the signal inputs to define the CMRR-related interaction between output voltage and output current, or the output impedance, for the improved transconductance amplifier.

where $CMRR_{DA}$ is the common-mode rejection of the difference amplifier, as analyzed in Sec. 1.3.2. Through this R_S term, the output impedance competes with the previously described output voltage swing, bandwidth, output offset, and input impedance in the design selection of that resistance.

References

1. J. Graeme, *Photodiode Amplifiers; Op Amp Solutions,* McGraw-Hill, New York, 1996.
2. Graeme, op cit., 1996.
3. J. Graeme, *Optimizing Op Amp Performance,* McGraw-Hill, New York, 1997.
4. Graeme, op cit., 1997.
5. Graeme, op cit., 1996.
6. Graeme, op cit., 1996.
7. Graeme, op cit., 1997.
8. G. Tobey, J. Graeme, and L. Huelsman, *Operational Amplifiers; Design and Applications,* McGraw-Hill, New York, 1971.
9. Tobey et al., op cit., 1971.
10. Graeme, op cit., 1996.
11. Graeme, op cit., 1996.
12. D. A. Bell, *Noise and the Solid State,* Wiley, New York, 1985.
13. Graeme, op cit., 1996.
14. Graeme, op cit., 1997.
15. Graeme, op cit., 1997.
16. Graeme, op cit., 1997.
17. Graeme, op cit., 1997.
18. Graeme, op cit., 1997.

5

Composite Amplifiers

Most op amp applications enclose only one amplifier within a given feedback loop but composite amplifiers enclose two. The two amplifiers connect in series, producing increased gain, accuracy, and bandwidth. This connection boosts the circuit's open-loop gain to the product of the individual amplifier gains. Those individual gains may be either open-loop or closed-loop, depending on the presence of local feedback about a given amplifier. In any case, the composite connection reduces the error contributions of both op amps and extends the circuit bandwidth.[1] This connection generally removes the significance of all of the second amplifier's input errors because these errors reflect to the composite input divided by the gain contribution of the first amplifier. Also, the composite connection reduces the gain and slewing errors of the first amplifier. To do so, the second op amp amplifies the output signal of the first, reducing the output signal swing required from the first op amp. With reduced output swing, the first op amp reflects less gain error to its inputs and encounters less output slewing demand. At higher frequencies, the continuing gain boost of the composite connection also extends the circuit bandwidth. As described in another publication,[2] the fundamental bandwidth limit of an op amp circuit occurs where the feedback demand for gain exceeds the available open-loop gain. By boosting the available gain, composite connections extend the bandwidth.

However, the two amplifiers in the same feedback loop produce a two-pole composite response, requiring phase compensation.[3] Numerous phase compensation methods address this requirement throughout this chapter. Most of these follow traditional phase compensation philosophy, modifying the A_{OL} response, but others modify the $1/\beta$ feedback response instead.[4] The majority tailor the composite open-loop response to restore a single-pole A_{OL} roll off around the $1/\beta$ feedback intercept. To do so, those compensation methods apply local feedback around one of the composite circuit's op amps, modifying that op amp's gain contribution to the composite response. This local feedback only affects the composite open-loop response, leaving the control of the closed-loop response to the composite feedback.

The local feedback described here independently configures one of the op amps as either a voltage amplifier, an integrator, or a differentiator. With the first compensation method, resistive voltage-amplifier feedback simply limits the second amplifier's gain contribution to the composite response. This moves the second amplifier's pole to a higher frequency, restoring a single-pole composite roll at lower frequencies. However, this simple approach limits the composite's loop-gain benefit at those lower frequencies. The second compensation method replaces the resistive solution with integrator feedback, restoring the lower-frequency loop gain. These first two feedback connections do not suit lower-gain applications. However, a modification of the integrator feedback method extends its application to all gain levels in a third-phase compensation case. With the fourth and fifth compensation methods, differentiator local feedback around first one and then the other op amp increases the bandwidth for lower-gain applications.

Departing from tradition, the sixth-phase compensation method results from a closer examination of the rate-of-closure stability criterion. This criterion defines frequency stability from the difference in slopes of the A_{OL} and $1/\beta$ curves at their intercept. The five other compensation methods control this difference primarily by shaping the composite A_{OL} response. The sixth method shapes the composite $1/\beta$ response instead. With the latter approach, the feedback-factor compensation produces the greatest loop gain improvement of all the compensation methods, and it does so with just one phase compensation element. However, this method requires more precise compensation selection and only serves higher-gain applications.

Most often, phase compensation selection, and the resulting performance evaluation, require individual response analyses of a specific op amp's configuration. However, one-time analyses of the six compensation methods just summarized yield standard design equations that replace the individual analyses for most composite amplifiers. Then, the selection of a given compensation method identifies a set of design equations that define component values and performance results.

5.1   Composite Amplifier Benefits

Composite amplifiers improve op amp circuit performance through reduced error and increased bandwidth. Both op amps of the composite structure reduce each other's error effects, producing greater accuracy than either amplifier produces when used alone. This symbiotic relationship results from the gain isolation provided by each op amp to the other. The gain of the composite's first op amp isolates the composite input from the input errors of the second op amp. Similarly, the gain of the second op amp isolates the first op amp's output from the larger signal swing of the composite output. Composite amplifiers boost circuit bandwidth by increasing the circuit's gain–bandwidth product. In the composite connection, the gains of the two op amps combine at all frequencies. Prior to the op amps' unity-gain crossovers f_{c1} and f_{c2}, this combination increases the net open-loop gain of the composite circuit. Increased open-loop gain at higher frequencies corresponds to a greater gain–bandwidth product. Thus at frequencies up to f_{c1} and f_{c2}, the composite circuit increases the bandwidth available at any gain level greater than unity.

5.1.1 Reducing errors with composite amplifiers

Figure 5.1 shows the basic composite amplifier, having two op amps connected in a common feedback loop. Feedback elements Z_1 and Z_2 supply the composite feedback and A_1 and A_2 serve as gain blocks within the loop. The series connection of the amplifiers boosts the composite open-loop gain to the product of the two amplifier gains, $A_{OLc} = A_{OL1}A_{OL2}$, and as will be described, this boosted gain also extends the bandwidth. At first, the same gain and bandwidth improvements would seem to result from a more straightforward approach. Simply

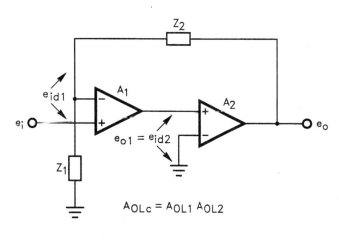

$$e_{id1} = V_{OS1} + e_{n1} + \frac{e_{id2}}{A_{OL1}} + \frac{e_i}{CMRR_1} + \frac{\delta V_{S1}}{PSRR_1}$$

Figure 5.1 Composite amplifiers connect two op amps in series within a common feedback loop for improved accuracy and speed.

connecting two op amps in series with individual feedback networks also improves gain accuracy and bandwidth over that attainable with a single amplifier. This connection divides the overall circuit's gain requirement into two lesser closed-loop gains. Then, the reduced gain demands of the two amplifiers increase the amplifiers' loop gains and bandwidths beyond those of a single amplifier solution.

However, this straightforward alternative lacks the major accuracy improvement afforded by the composite connection. In the straightforward case, the error sources of both op amps contribute directly to the overall circuit error. In the composite case, the two op amps in the same feedback loop reduce the error effects of both amplifiers significantly. Examination of the amplifiers' input error signals e_{id1} and e_{id2} in Fig 5.1 reveals this error reduction. The net differential-input error signal of an op amp combines numerous effects, as expressed by[5]

$$e_{id} = V_{OS} + e_n + \frac{e_o}{A_{OL}} + \frac{e_{icm}}{CMRR} + \frac{\delta V_S}{PSRR}$$

In this equation, the first two e_{id} terms represent the op amp's net input offset voltage and input noise voltage. For simplicity, the V_{OS} term here lumps the input offset effects of amplifier input bias currents with the op amp's input offset voltage. The last three terms of the e_{id} equation represent the input error signals resulting from finite values of open-loop gain, common-mode rejection, and power-supply rejection. The gain error signal e_o/A_{OL} typifies these latter errors. Finite A_{OL} gain requires an input signal of e_o/A_{OL} to support an output signal e_o. This input signal represents a departure from the ideal, infinite gain case, making e_o/A_{OL} an error signal.

Referring to Fig. 5.1, e_{id1} appears directly at the composite amplifier input where it defines the input errors for the overall circuit. The circuit amplifies e_{id1} by a gain of $1/\beta_c$, where β_c is the feedback factor of the composite circuit, or $\beta_c = Z_1/(Z_1 + Z_2)$.[6] Together, e_{id1} and β_c define the composite circuit's output error as e_{id1}/β_c. The second amplifier's input error signal e_{id2} also influences this output error but only through its influence upon e_{id1}. In the composite connection, $e_{o1} = e_{id2}$, and this reflects to e_{id1} through A_1's gain error term, $e_{o1}/A_{OL1} = e_{id2}/A_{OL1}$. Examination of the basic e_{id} equation shows no other mechanism by which e_{id2} influences e_{id1}. Thus the e_{id2} input errors of A_2 reflect to the composite circuit input attenuated by the open-loop gain A_{OL1}. This virtually eliminates the effects of A_2's input errors in the composite connection. Further, the composite circuit makes e_{o1} a very small signal, reducing the gain error signal of the first op amp. Without the second op amp, the first amplifier would support the full output signal e_o, resulting in an input gain error signal of e_o/A_{OL1}. With the second amplifier, e_{o1} only supports a signal of e_o/A_{OL2}, reducing the composite input gain error to $e_o/A_{OL1}A_{OL2}$. Thus the composite connection reduces the gain error signal of the first op amp by a factor of A_{OL2}.

These error reductions greatly improve circuit accuracy when using a specialized, but less accurate second op amp. For example, high-power and high-speed op amps deliver their special capabilities at the composite output without introducing significant errors in the composite response. Choosing A_1 for input accuracy minimizes the composite error and choosing A_2 for its specialty optimizes load driving and slewing performance. For high-power applications, high-level output currents delivered by A_2 produce thermal feedback inside this amplifier, modulating the A_2 input offset voltage. The resulting offset change δV_{OS2} appears in e_{id2} and reflects to the composite circuit input as a small $\delta V_{OS2}/A_{OL1}$. Similarly, in high-speed applications, high output slew rates place little demand on the input amplifier. The composite structure reduces the A_1 output swing to just e_{id2}. This signal reflects output slewing demands to A_1 through the gain error signal of A_2, which equals e_o/A_{OL2}. Thus an output slew rate of $\delta e_o/\delta t$ requires an A_1 output slew of only $(\delta e_o/\delta t)/A_{OL2}$. In practice, most composite amplifier phase compensation methods reduce the A_2 gain isolation from A_{OL2} to a closed-loop A_{CL2}. Still, the reduction in A_1 slew demand generally remains large.

A common op amp test loop[7] also benefits from the reduced e_{o1} output swing. This loop uses the composite structure to isolate the amplifier under test from large test signals. Basically, the loop consists of the circuit in Fig. 5.1 along with one of several phase compensation networks described later. In this loop, A_1 becomes the amplifier tested and A_2 reduces signal swing at the A_1 output. This signal reduction aids in the testing of amplifier characteristics such as common-mode rejection and power-supply rejection. For these tests, the composite circuit amplifies the A_1 input error signal, producing an easily measurable output signal e_o. However, only a small fraction of this signal, e_o/A_{OL2}, appears at the A_1 output. Without the composite structure, the A_1 output supports the full e_o signal. Then the finite open-loop gain of A_1 develops an input error signal of e_o/A_{OL1}, and this error signal adds to the one being measured, confounding the measurement. With the composite, the corresponding input error signal drops to $e_o/A_{OL2}A_{OL1}$, removing the significance of the A_1 gain error in the measurement.

5.1.2 Increasing bandwidth with composite amplifiers

The gain boost of the composite structure extends the available circuit bandwidth for almost all op amp applications. Some applications limit bandwidth intentionally through the choice of feedback elements, but the circuit's available bandwidth remains greater with the composite structure. Figure 5.2 demonstrates this improvement with

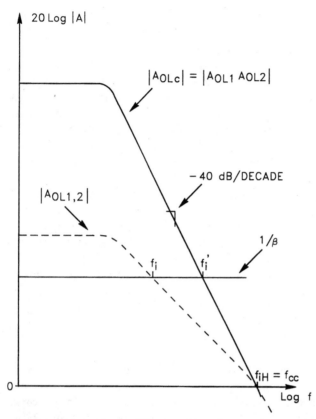

Figure 5.2 Composite amplifiers combine the open-loop gains of two op amps into a composite A_{OLc}, increasing loop gain and bandwidth.

the A_{OL} and $1/\beta$ curves relevant to the discussion that follows. In the figure, A_{OL1} or A_{OL2} represents the open-loop response of the individual op amps. For simplicity, the example shown uses the same response plot for the two amplifiers, indicating the same amplifier type. Using different types for the two produces different response curves and later discussions account for this difference. The composite response shown combines the two op amp responses, increasing the net open-loop gain to $A_{OLc} = A_{OL1}A_{OL2}$. On the logarithmic axes of the plots, linear summation of the A_{OL1} and A_{OL2} responses produces the composite gain A_{OLc}. Visual comparison of the A_{OL1} and A_{OLc} responses reveals the gain–bandwidth product improvement produced by the composite amplifier.

Adding the circuit's $1/\beta$ curve to the figure provides a more quantitative evaluation of this improvement. That curve represents the feedback demand for open-loop gain, and the A_{OL} curve represents the circuit's ability to supply this demand. Where demand exceeds supply, a circuit's closed-loop response rolls off in a bandwidth limit. Intercepts of the $1/\beta$ and A_{OL} curves mark the transitions between satisfied and unsatisfied gain demands. First, consider the $1/\beta$ curve shown for the common voltage-amplifier op amp connection. There, a flat response curve results from simple resistive feedback, reflecting a frequency-independent feedback demand for gain. Both the single amplifier and the composite amplifier responses shown meet this demand at lower frequencies, where the $1/\beta$ curve remains within the boundaries of both A_{OL1} and A_{OLc}. However, at higher frequencies, the $1/\beta$ curve crosses both gain responses, entering regions of unsupported feedback demand. For the single-amplifier case, $1/\beta$ makes this transition at its f_i intercept with A_{OL1}, giving $BW = f_i$. The composite A_{OLc} response extends this bandwidth-defining intercept to the higher frequency f_i'. In practice, phase compensation reduces the f_i' improvement for most composite amplifiers. Still, these curves display the maximum improvement potential, and the last phase compensation method described in this chapter achieves this full potential.

The bandwidth improvement described here focuses upon the more common objective of extending high-frequency gain. However, the composite amplifier also expands the bandwidth by increasing low-frequency gain in the case of the common integrator circuit. As shown in Fig. 5.3, this circuit produces an ideal closed-loop gain of $-1/RCs$. Here, the s term of the denominator makes the ideal gain rise with decreasing frequency, approaching infinity as frequency drops toward zero. However, the finite open-loop gain of the op amp fails to support such a closed-loop gain, resulting in a second bandwidth limit for the practical integrator. This added limit defines a minimum rather than maximum response frequency in a low-frequency bandwidth limit.

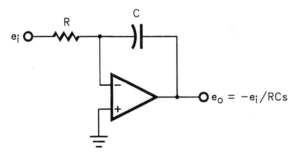

Figure 5.3 The common integrator circuit produces an ideal closed-loop gain $-1/RCs$ that approaches infinity as frequency approaches zero and develops a second bandwidth limit.

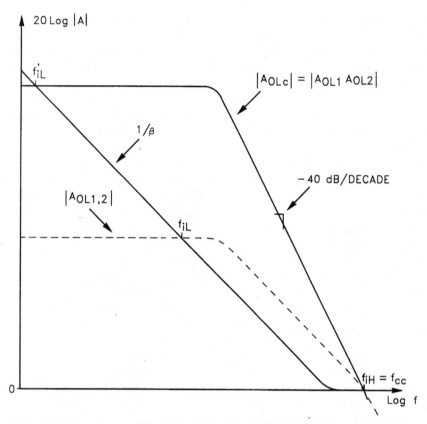

Figure 5.4 Composite amplifiers extend the integrator's frequency range by moving its low-frequency A_{OL} and $1/\beta$ intercept from f_{iL} down to f'_{iL}.

Replacing the common integrator's single op amp with the two of the composite connection moves the latter limit to a much lower frequency.

Figure 5.4 demonstrates this low-frequency limit for the single op amp and for the composite amplifier. The $1/\beta$ response of the integrator intercepts both the single and the composite A_{OL} responses twice, producing dual bandwidth limits in each case. Examination of the integrator circuit in Fig. 5.3 reveals this circuit's $1/\beta$ response through the voltage divider ratio of the feedback network which defines the feedback factor $\beta = RCs/(1 + RCs)$. Then $1/\beta = 1 + 1/RCs$, as reflected by the $1/\beta$ curve of Fig. 5.4. As with other op amp connections, the integrator's $1/\beta$ curve encounters a bandwidth limit imposed by this curve's high-frequency intercept with the A_{OL} responses. At higher frequencies, the integrator's $1/\beta$ curve levels off at the unity-gain axis and intercepts both A_{OL} responses of the example at $f_{iH} = f_{cc}$. Although unusual, the integrator's $1/\beta$ curve also produces a second intercept at low frequencies. There, the integrator's increasing gain demands make the $1/\beta$ curve intercept both A_{OL} responses a second time. These intercepts mark a second departure between the available gain and the feedback demand. Below these intercept frequencies, the integrator response also rolls off. For the single-amplifier integrator, limited open-loop gain places this low-frequency roll off at f_{iL} in Fig. 5.4. The increased gain of the composite A_{OLc} moves this roll off to the much lower frequency f'_{iL}.

In most cases, composite amplifiers increase the available open-loop gain sufficiently to remove the practical significance of the integrator's low-frequency intercept. The composite's added gain moves this intercept to such a high level that noise rather than response roll off dominates circuit error. There, the very high integrator gain directly amplifies the amplifier input noise voltage, making the roll-off error secondary. However, at frequencies above this new limit, the added open-loop gain of the composite amplifier extends the integrator's useful low-frequency response significantly.

While Fig. 5.4 serves to illustrate the composite amplifier bandwidth improvement qualitatively, the response curves shown there do not represent a practical implementation. Practical composite amplifiers require phase compensation that alters these curves, generally reducing the degree of performance improvement. Linear summation of the A_{OL1} and A_{OL2} curves, to define A_{OLc}, also transfers the individual response poles of A_{OL1} and A_{OL2} to A_{OLc}. The resulting two-pole A_{OLc} response corresponds to 180° of phase shift through the composite amplifier. Placing this amplifier in a feedback loop automatically inserts 180° of feedback phase shift, inviting oscillation, and phase compensation must be added to make the composite amplifier a practical solution. The remaining sections of this chapter describe six phase compensation methods for composite amplifiers, including a compensation modification specifically serving the integrator example described here.

5.2   Gain-Limiting Phase Compensation

The simplest composite-amplifier phase compensation moves one of the circuit's two poles to a higher frequency just by limiting the gain supplied by A_2. In the uncompensated state, amplifier A_2 contributes its entire open-loop gain to the composite open-loop gain. Unfortunately, A_2 also transfers its low-frequency open-loop response pole to the composite open-loop response. Local feedback around A_2, and inside the composite loop, reduces this amplifier's gain contribution but moves the associated pole to a higher frequency. Trading gain for bandwidth, the gain-limiting compensation produces a region of single-pole roll off in the composite open-loop response. In this region, the phase shift of the composite amplifier drops from the original 180° to

90°, permitting stable feedback connections. However, this single-pole response only includes the higher-gain regions of the composite response, restricting the composite circuit operation to such gains. With this compensation method, further reduction in the gain contribution of A_2 extends the single-pole region to lower gains but simultaneously reduces the gain and bandwidth benefits of the composite circuit. Thus a compromise guides the selection of A_2's closed-loop gain based on the closed-loop gain required of the composite circuit. Analysis yields a simple design equation relating the two closed-loop gains.

5.2.1 Adding gain-limiting compensation

Figure 5.5 illustrates this case with local feedback connected around A_2. In the figure, resistors R_1 and R_2 supply the composite circuit's feedback, and the addition of R_3 and R_4 produces the local gain-limiting feedback. With the added feedback, the noninverting amplifier formed by A_2, R_3, and R_4 boosts the effective open-loop gain of the composite circuit by a controlled amount. Previously, the basic composite amplifier produced a net open-loop gain of $A_{OLc} = A_{OL1}A_{OL2}$, where A_{OL1} and A_{OL2} represent the open-loop gains of the individual op amps. Figure 5.5 reduces the net gain to $A_{OLc} = A_{OL1}A_{CL2}$, where A_{CL2} is the closed-loop gain of A_2.

Response plot analysis[8] demonstrates the benefits of Fig. 5.5 and the

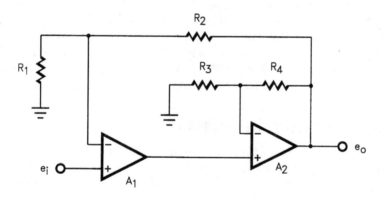

$$1 + \frac{R_4}{R_3} = \sqrt{\frac{f_{c2}}{f_{c1}}\left(1 + \frac{R_2}{R_1}\right)} \qquad\qquad e_o = \left(1 + \frac{R_2}{R_1}\right)e_i$$

Figure 5.5 Separate local feedback limits the gain provided by A_2 to move this amplifier's pole to a higher frequency and establish frequency stability.

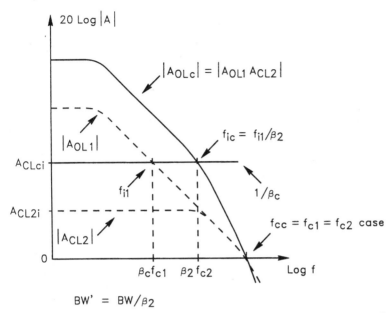

$$BW' = BW/\beta_2$$

Figure 5.6 Optimum selection of local feedback for Fig. 5.5 places the second A_{OLc} pole at the f_{ic} intercept for 45° of phase margin and maximum bandwidth.

stability limitation in Fig. 5.6. There, the composite gain curve A_{OLc} results from its component gain curves A_{OL1} and A_{CL2}. On the logarithmic gain scale, linear summation of the component curves produces the net A_{OLc} curve. For simplicity, the example shown illustrates the common dual-amplifier case with equal unity-gain crossover frequencies, $f_{c1} = f_{c2}$, for the two op amps. Different amplifiers in the composite's two positions produce different response curves. However, the design equations derived next adjust for this difference. For the example case here, linear summation of the component curves in Fig. 5.6 leaves the crossover frequency and for the composite A_{OLc}, $f_{cc} = f_{c1} = f_{c2}$.

Comparing the A_{OLc} curve with the single-amplifier A_{OL1} curve demonstrates the gain and bandwidth improvements of the composite circuit. The space separating these curves shows that the composite A_{OLc} increases both loop gain and closed-loop bandwidth. Loop gain equals the vertical distance from the $1/\beta_c$ curve up to the open-loop response. Up to the f_{ic} intercept, moving the net A_{OL} from A_{OL1} to A_{OLc} increases this distance, and loop gain, by a factor equal to A_{CL2}. Similarly, the A_{OL} move increases the closed-loop bandwidth. This bandwidth equals the frequency of the $1/\beta_c$ intercept with the open-loop response, and the composite amplifier moves this limit from the f_{i1} intercept with A_{OL1} to the f_{ic} intercept with A_{OLc}.

As expected, this curve summation still transfers the closed-loop response pole of A_{CL2} to A_{OLc}. There, the added pole results in a two-pole, open-loop response, suggesting continued instability potential and guiding the selection of A_{CL2}. This pole occurs at a frequency of $\beta_2 f_{c2}$, where $\beta_2 = R_3/(R_3 + R_4)$ is the feedback factor of the A_2 local feedback and f_{c2} is the unity-gain crossover of A_{CL2}. Note that the noninverting amplifier formed with A_2 produces an ideal gain of $A_{CL2i} = 1 + R_4/R_3 = 1/\beta_2$. This makes the frequency of the second pole, $\beta_2 f_{c2} = f_{c2}/A_{CL2i}$. Thus increasing the A_{CL2i} gain boost moves the second pole down in frequency, encompassing more of the A_{OLc} response. However, this second pole restricts the use of the composite amplifier to lower frequencies and higher circuit gains. Frequency stability requires a $1/\beta_c$ intercept with A_{OLc} before the roll off of the second pole fully develops. Higher composite circuit gain A_{CLci} raises the $1/\beta_c$ curve for an intercept at f_{ic} in a region of reduced A_{OLc} slope. There, the composite circuit feedback defines $1/\beta_c = 1 + R_2/R_1 = A_{CLci}$, where A_{CLci} is the ideal closed-loop gain of the composite circuit. In the design process, the application requirement determines A_{CLci} and, thereby, $1/\beta_c$. To ensure stability, the choice of A_{CL2i} then sets an appropriate location for the second pole.

5.2.2 Optimizing the gain-limiting compensation

Several relationships between the response plots in Fig. 5.6 quantify the bandwidth improvement of the composite amplifier and guide the choice of A_{CL2i}. First, note the relationship between the A_{CL2i} gain boost and the bandwidth improvement from f_{i1} to f_{ic}. Preceding f_{ic}, both the A_{OL1} and the A_{OLc} curves follow the 1:1 gain–frequency slopes of single-pole responses. There, the composite gain boost of A_{CL2i} reflects through a 1:1 slope for an equal increase in the $1/\beta_c$ intercept frequency. Thus $f_{ic} = A_{CL2i}f_{i1} = f_{i1}/\beta_2$, and the composite connection increases the bandwidth by a factor equaling $A_{CL2i} = 1/\beta_2$.

Next, stability requirements align the second pole with f_{ic}, producing a second equation for this frequency. Then, simultaneous solution of the two f_{ic} equations defines the optimum A_{CL2i}. For optimum gain–bandwidth product, A_{CL2i} is increased until the second A_{OLc} pole retreats to the $1/\beta_c$ intercept. This pole placement adds 45° of feedback phase shift to the 90° from the first A_{OLc} pole, setting the phase margin at $\phi_m = 180° - (45° + 90°) = 45°$. As mentioned earlier, the second pole occurs at $\beta_2 f_{c2}$ and the pole placement described here makes $\beta_2 f_{c2} = f_{ic}$. But from the last paragraph, $f_{ic} = f_{i1}/\beta_2$, and solving for β_2 yields $\beta_2 = \sqrt{f_{i1}/f_{c2}}$. From the curves, $f_{i1} = \beta_c f_{c1}$, making $\beta_2 = \sqrt{\beta_c f_{c1}/f_{c2}}$. Then from before, $\beta_2 = R_3/(R_3 + R_4) = 1/A_{CL2i}$ and $\beta_c = R_1/(R_1 + R_2)$. Solving for A_{CL2i} yields the final design equation

$$A_{\mathrm{CL}2i} = 1 + \frac{R_4}{R_3} = \sqrt{\frac{f_{c2}}{f_{c1}}\left(1 + \frac{R_2}{R_1}\right)}$$

Then, for a given pair of amplifier crossover frequencies f_{c1} and f_{c2}, the choice of R_2/R_1 to set the composite closed-loop gain also defines R_4/R_3 to set the local feedback gain.

The $A_{\mathrm{CL}2i}$ equation defines a maximum gain for which the phase margin remains at least 45°. This gain limit indirectly imposes a greater bandwidth requirement upon A_2 for lower-gain applications. Substituting $A_{\mathrm{CL}ci} = 1 + R_2/R_1$ in this equation defines the relationship between the circuit's two closed-loop gains as $A_{\mathrm{CL}2i} = \sqrt{(f_{c2}/f_{c1})A_{\mathrm{CL}ci}}$. Thus the ideal closed-loop gain $A_{\mathrm{CL}ci}$ of the composite circuit defines a limit for the gain boost permitted from $A_{\mathrm{CL}2i}$. For the common case of $f_{c2} = f_{c1}$, $A_{\mathrm{CL}2i} = \sqrt{A_{\mathrm{CL}ci}}$ and the two gains approach unity together. Yet the benefit of a composite amplifier results only from the gain boost produced by A_2. Thus lower values of $A_{\mathrm{CL}ci}$ require that $f_{c2} \gg f_{c1}$ to retain a beneficial level of $A_{\mathrm{CL}2i}$.

5.3 Integrator-Feedback Phase Compensation

The gain-limiting compensation of the preceding section sacrifices open-loop gain for phase compensation simplicity. There, gain limiting reduces $A_{\mathrm{OL}c}$ from $A_{\mathrm{OL}1}A_{\mathrm{OL}2}$ to $A_{\mathrm{OL}1}A_{\mathrm{CL}2}$ across the entire frequency range. For stability, this compensation method must sacrifice gain at frequencies surrounding the $1/\beta_c$ intercept. However, below this region, the gain reduction serves no purpose. At these lower frequencies, integrator rather than resistive feedback around A_2 restores the lower-frequency gain boost without compromising stability. This phase compensation alternative requires one additional phase compensation element and, once again, does not serve applications having lower closed-loop gains. Later an alternative integrator-feedback compensation method extends this approach to those lower gains.

5.3.1 Adding integrator-feedback compensation

Figure 5.7 shows three modifications to the previous composite amplifier that introduce the integrator-feedback compensation. There, integrating capacitor C_1 now blocks low-frequency feedback around A_2, amplifier A_2 switches to the inverting configuration, and the composite circuit feedback now drives A_1's noninverting input. The capacitor and the inverting configuration convert A_2 to a modified integrator, and at lower frequencies, A_2 responds as a true integrator. There it contributes a greater closed-loop gain inside the composite feedback loop.

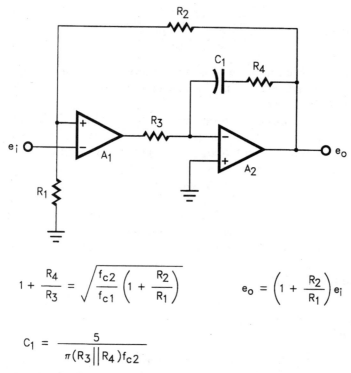

$$1 + \frac{R_4}{R_3} = \sqrt{\frac{f_{c2}}{f_{c1}}\left(1 + \frac{R_2}{R_1}\right)} \qquad\qquad e_o = \left(1 + \frac{R_2}{R_1}\right)e_i$$

$$C_1 = \frac{5}{\pi(R_3 \| R_4)f_{c2}}$$

Figure 5.7 Feedback-blocking capacitor C_1 forms a modified integrator to increase the gain contributed by A_2 at lower frequencies, while retaining the frequency stability conditions of Fig. 5.5.

At higher frequencies, resistor R_4 overrides the integrator feedback, returning A_2 to a voltage amplifier and repeating the stability control of the preceding gain-limiting case. Switching A_2 to the inverting configuration, to form the modified integrator, also inverts the polarity of this amplifier's gain. This requires switching the composite feedback connected to A_1 to that amplifier's noninverting input.

Figure 5.8 shows the resulting response curves, which produce the low-frequency gain boost through the integrator local feedback. In the figure, A_{CL2} follows an integrator response at low frequencies and then returns to a voltage-amplifier response at higher frequencies. Linear summation of the A_{CL2} and A_{OL1} curves again produces the composite open-loop gain curve A_{OLc}. Due to the A_{CL2} integrator response, A_{OLc} achieves much higher gain levels at lower frequencies. At higher frequencies, the A_{CL2} voltage amplifier response retains the frequency stability characteristics described for the gain-limiting case. For simplicity, the example curves shown represent the $f_{c1} = f_{c2}$ condition encountered when using the same amplifier type for the circuit's two op amps.

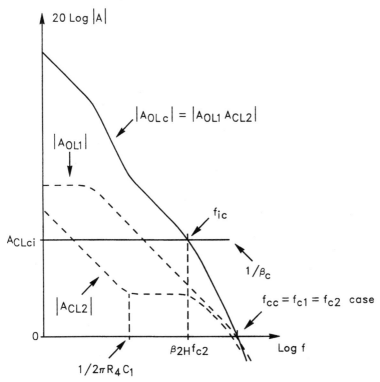

Figure 5.8 The blocking capacitor of Fig. 5.6 greatly increases the low-frequency A_{OLc} gain without disturbing the stability conditions at f_{ic}.

However, the design equations and discussion that follow accommodate the more general case of different amplifier types.

5.3.2 Optimizing the integrator-feedback compensation

Closer examination of Fig. 5.8 defines the phase compensation design equations. As in the gain-limiting case before, placing the second pole of A_{OLc} at the intercept frequency f_{ic} retains 45° of phase margin without unnecessarily restricting bandwidth. The intercept of the $1/\beta_c$ curve with A_{OLc} defines f_{ic} and graphical analysis quantifies the result. As shown, f_{ic} marks the beginning of a region where both A_{OL1} and A_{CL2} follow single-pole roll offs. In that region, $A_{OL1} \approx f_{c1}/f$ and $A_{CL2} \approx f_{c2}/f$, making $A_{OLc} = A_{OL1}A_{CL2} \approx f_{c1} f_{c2}/f^2$. At the intercept, $f = f_{ic}$, making $1/\beta_c = A_{OLc} \approx f_{c1}f_{c2}/f_{ic}^2$. Solving for f_{ic} produces $f_{ic} = \sqrt{\beta_c f_{c1} f_{c2}}$, where $\beta_c = R_1/(R_1 + R_2)$. The second A_{OLc} pole results from the high-frequency roll off of A_{CL2} and occurs at $\beta_{2H}f_{c2}$, where β_{2H} is the high-frequency value of β_2. At high frequencies, integrator capacitor C_1 of the circuit does not affect β_2 and $\beta_{2H} = R_3/(R_3 + R_4)$. Thus $f_{ic} = f_{c2}R_3/(R_3 + R_4)$, just as in the preceding gain-limiting case, and equating this result with $f_{ic} = \sqrt{\beta_c f_{c1} f_{c2}}$ yields the design equation for R_3 and R_4,

$$1 + \frac{R_4}{R_3} = \sqrt{\frac{f_{c2}}{f_{c1}}\left(1 + \frac{R_2}{R_1}\right)}$$

Thus knowing f_{c2} and f_{c1} and selecting R_1 and R_2 to set the composite gain once again defines R_4/R_3.

Previously, this selection of resistors set the circuit's phase margin at 45°. Preservation of this phase margin guides the selection of the integrator capacitor here. This selection defines the A_2 transition between integrator and voltage-amplifier functions at $1/2\pi R_4 C_1$. As shown in Fig. 5.8, the transition produces a response zero for A_{OLc}, reducing its response slope for a stable intercept at f_{ic}. To be effective, the region of reduced slope must span at least a decade of frequency prior to f_{ic}. This span develops the full phase benefit of the response zero before the f_{ic} intercept occurs. A greater frequency span would unnecessarily reduce the A_{OLc} benefit produced by the integrator response. In compromise, setting $1/2\pi R_4 C_1 = f_{ic}/10 = \beta_{2H}f_{c2}/10$ and solving produces the C_1 design equation

$$C_1 = \frac{5}{(R_3 \| R_4)\pi f_{c2}}$$

The integrator-feedback phase compensation described here only serves composite amplifiers having closed-loop gains of 3 or greater. At gains lower than this, the second pole of each op amp can introduce significant phase shift in the composite loop. These second poles occur around the unity-gain crossover frequencies of the op amps, producing high phase shift for A_{OLc} at its unity-gain crossover f_{cc}. Lower closed-loop gain corresponds to a $1/\beta$ curve near the unity-gain axis, placing the f_{ic} intercept close to f_{cc}. This placement makes the high A_{OLc} phase at f_{cc} a significant contributor to feedback phase shift. Normally, this requires that $1/\beta \geq 3$ to limit this effect. Alternately, feedback factor modification adjusts the composite amplifier phase compensation as described in the next section.

5.3.3 Alternative integrator-feedback compensation

Feedback factor modification extends the integrator-feedback compensation of composite amplifiers to lower gain applications. In this case, changes in feedback elements raise the circuit's $1/\beta$ curve without changing its closed-loop gain. Raising the $1/\beta$ curve moves the $1/\beta$ intercept with A_{OLc} up and away from the high phase shift at f_{cc}. For stability, the result remains the same as if the composite amplifier were connected for a higher closed-loop gain.

The integrator function serves to illustrate this approach, and connecting the composite amplifier in this configuration places the maximum stability demand upon the circuit. At high frequencies, the basic integrator's $1/\beta$ curve drops to the unity-gain axis and intercepts A_{OL} at the unity-gain crossover frequency f_c, as shown previously in Fig. 5.4. For the composite amplifier, this places the intercept at f_{cc}, inserting a high level of phase shift in the loop response. This would normally make the composite amplifier unsuitable for the integrator function. Still, the composite amplifier offers great promise to this function, as described in Sec. 5.1.2. There, the added low-frequency gain of the composite amplifier greatly extends the useful bandwidth of the integrator circuit.

To realize this benefit, Fig. 5.9 combines the integrator-feedback phase compensation of the preceding section with feedback-factor phase compensation. Later, Sec. 5.5 develops the latter compensation for more general applications. In Fig. 5.9, capacitor C_3 provides the phase compensation adjustment. Without C_3, the composite amplifier feedback produces a feedback factor of $R_1C_1s/(1 + 1/R_1C_1s)$, and this feedback factor reduces to unity at high frequencies. With C_3, the high-frequency feedback factor becomes $\beta_{cH} = C_1/(C_1 + C_3)$, making $\beta_{cH} < 1$. The selection of C_3 adjusts the composite circuit's high-frequency feedback factor to any desired level prior to the stability-defining intercept at f_{ic}. Further, the presence of C_3 does not alter the circuit's closed-loop response. Input signal e_i only drives R_1 to produce the cir-

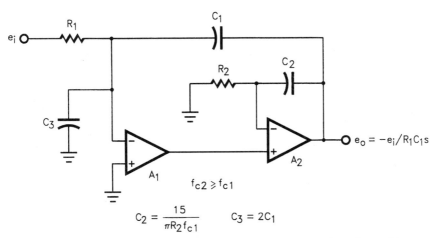

Figure 5.9 Adding C_3 to the integrator-feedback phase compensation adapts this compensation method to a composite integrator function.

cuit's feedback current and, thereby, the circuit's ideal response remains the $e_o = -e_i/R_1C_1s$ of the basic integrator.

However, the presence of C_3 does provide a capacitive return to ground for C_1, placing a capacitive load of $C_1C_3/(C_1 + C_3)$ on the composite amplifier output. Such capacitance loading can degrade op amp stability. However, the loading here only becomes significant for very low-gain integrators that could suggest large capacitance values for C_1. In most cases, simply increasing R_1 instead of C_1 produces the desired low gain without producing excessive capacitance loading.

5.3.4 Optimizing the alternative integrator compensation

Figure 5.10 shows the response curves that guide the component selection for this phase compensation. There, A_{CL2} displays a modified integrator response, leveling off at the unity-gain axis. This leveling results from the noninverting connection of the A_2 integrator, which now produces a closed-loop response of $1 + 1/R_2C_2s$. Linearly adding the A_{CL2} curve to the A_{OL1} curve again produces the composite open-loop curve A_{OLc}. There, A_{OLc} displays a response zero produced by the leveling of the A_{CL2} response. Following this zero, A_{OLc} displays a region of single-pole roll off, suitable for a stable $1/\beta$ intercept. As desired, the composite integrator's $1/\beta$ curve, $1/\beta_c$, intercepts A_{OLc} in this single-pole region at level $1/\beta_{cH}$, away from f_{cc}.

The stability requirements of this intercept define the design equations for the phase compensation elements. First, a β_{cH} requirement defines C_3 to limit the phase effects of the op amps' second poles. Those poles occur around the op amps' unity-gain crossover frequencies at f_{c1} and f_{c2}. Selecting an A_2 op amp having $f_{c2} \gg f_{c1}$ would remove this amplifier's phase contribution at f_{cc}. However, an A_2 op amp with an f_{c2} as low as $f_{c2} = f_{c1}$ still serves this application, and the analysis that follows accommodates this worst-case condition. Under this $f_{c2} = f_{c1}$ condition, each op amp's second pole produces around 45° of additional phase shift at f_{cc}, raising the net A_{OLc} phase shift to 180°. Moving the intercept back to $f_{ic} = f_{cc}/3 = f_{c1}/3$ places the intercept at a point of reduced A_{OLc} phase shift. At $f_{cc}/3$, this phase shift drops to around 135°, leaving a 45° phase margin. In this single-pole region of A_{OLc}, the factor of three reduction in intercept frequency requires that $1/\beta_{cH} = 3$. From the earlier circuit discussion, $\beta_{cH} = C_1/(C_1 + C_3)$ and combining this with $1/\beta_{cH} = 3$ produces

$$C_3 = 2C_1$$

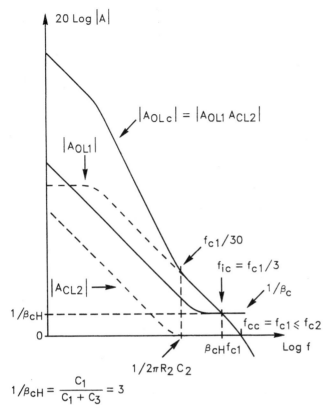

Figure 5.10 Capacitor C_3 of Fig. 5.9 levels off that integrator's $1/\beta_c$ curve above the unity-gain axis, moving the f_{ic} intercept away from the poor phase conditions at f_{cc}.

A second stability requirement defines the local feedback elements for A_2. This requirement places the A_{OLc} response zero well below f_{ic}, ensuring that the full benefit of the zero's phase correction exists at the f_{ic} intercept. The zero results from the $1/2\pi R_2 C_2$ break frequency of the A_2 local feedback, and placing the break at $f_{ic}/10$ provides the required separation from f_{ic}. From before, $f_{ic} = f_{c1}/3$, making $f_{ic}/10 = f_{c1}/30$, and equating $1/2\pi R_2 C_2 = f_{c1}/30$ yields

$$C_2 = \frac{15}{\pi R_2 f_{c1}}$$

To use this equation, first select a value for R_2 based on general resistance level considerations.

5.4   Differentiator-Feedback Phase Compensation

This phase compensation method specifically addresses the stability requirements of lower-gain applications. It lacks the low-frequency gain boost of the preceding two alternatives but extends the bandwidth further in lower-gain cases. Here, differentiator feedback around either op amp of the composite amplifier introduces an active zero to phase compensation. The more commonly used passive zero actually produces its response rise through signal attenuation at lower frequencies. Gain, rather than attenuation, produces the composite amplifier's active zero and the differentiator-connected op amp provides the supporting gain. This gain counteracts a circuit's response pole more effectively, extending the bandwidth of the composite circuit beyond that achieved by other means.

The examples considered here apply the active zero to compensate for an op amp's second pole. Lightly compensated op amps produce such a pole prior to their unity-gain crossover frequency. The lighter compensation improves the amplifier's slew rate but generally restricts the amplifier's use to higher-gain applications. In lower-gain applications, the second pole compromises stability, requiring phase compensation. Traditionally, passive compensation techniques address this second pole by modifying the circuit's feedback.[9] This largely avoids the second pole's phase effect at the $1/\beta$ intercept but leaves the bandwidth restricted to that pole's frequency. In this section, differentiator-feedback compensation of a composite amplifier compensates, rather than avoids, the effects of the second pole to extend the bandwidth.

5.4.1   Adding differentiator-feedback compensation

Adding one capacitor to the previous gain-limiting composite circuit converts that circuit for this new compensation. In Fig. 5.11, the capacitor shown converts the local feedback of A_2 to that of a modified differentiator. However, the A_2 amplifier's noninverting configuration and the presence of R_3 deviate from the ideal differentiator circuit. The noninverting configuration and the capacitor make A_2 perform as a voltage follower at lower frequencies. There, capacitor C blocks the feedback current, no signal develops on R_4, and the A_2 output follows its input. This follower action transmits the A_1 output signal to the composite output with unity gain. Thus at lower frequencies, the open-loop response of A_1 alone determines the composite open-loop response.

Choosing A_1 for unity-gain stability would then ensure a stable circuit. However, for the more lightly compensated example considered here, the open-loop response of A_1 exhibits a second pole that would increase the composite open-loop response slope to -40 dB per decade at higher frequencies. Such a slope would compromise stability for

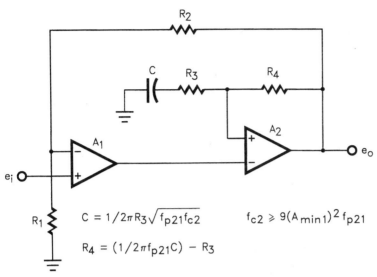

Figure 5.11 Differentiator feedback around A_2 introduces an active zero in the composite amplifier's feedback path to compensate for a second A_1 pole.

lower composite closed-loop gains. To compensate, the capacitor C conducts feedback current at higher frequencies, increasing the gain supplied by A_2. This gain continues to increase with frequency, producing an active zero that counteracts the roll off of A_1's second pole. However, the response rise of this zero would also produce instability with the ideal differentiator circuit, requiring the addition of resistor R_3 to limit the gain rise.[10]

Figure 5.12 illustrates the response correction of the circuit in Fig. 5.11. The A_{OL1} response rolls off with a second pole at f_{p21}, displaying the initial compensation requirement. Here, f_{p21} denotes the second pole of the first amplifier, A_1, and produces a stability-threatening -40-dB per decade roll off above that frequency. Normally, this response slope limits the amplifier's use to gains of A_{min1} or greater, where A_{min1} equals the value of A_{OL1} at f_{p21}. The traditional passive phase compensation methods described elsewhere[11] would extend the use to lower gains by raising the circuit's $1/\beta$ curve at high frequencies. This shifts the $1/\beta$ intercept to a region of single-pole A_{OL1} roll off. However, this also moves the intercept back in frequency, setting a bandwidth limit at f_{p21}.

To improve the bandwidth, the differentiator feedback compensation produces a composite A_{OLc} response that extends the region of single-pole roll off. Adding A_2 and its feedback introduces the compensating zero of the A_{CL2} response, as shown by a rising dashed line. Linear summation of the A_{OL1} and A_{CL2} curves produces the net A_{OLc}

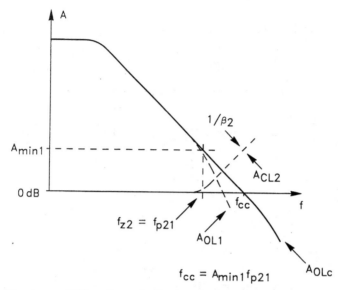

Figure 5.12 Differentiator feedback of Fig. 5.11 produces an A_{CL2} zero that compensates for the f_{p21} pole of A_{OL1}, extending the single-pole range of the composite A_{OLc}.

result, displaying the extended single-pole response. In the example shown, the zero of A_{CL2} extends the single-pole A_{OLc} roll off beyond the 0-dB or unity-gain crossing. This case permits stable operation for closed-loop gains all the way down to unity, and associated $1/\beta_c$ intercepts in the extended single-pole region define bandwidths potentially far greater than the previous f_{p21} limit. The maximum attainable bandwidth moves from f_{p21} to f_{cc}, where f_{cc} represents the unity-gain crossover frequency of the composite A_{OLc}.

5.4.2 Optimizing the differentiator-feedback compensation

Analysis of the response curves in Fig. 5.12 defines the design equations for the differentiator-feedback phase compensation. The single-pole roll off of the A_{OLc} response produces a simple relationship between f_{cc} and the previous f_{p21} bandwidth limit. This 20-dB per decade roll off produces a 1:1 slope for A_{OLc}. Thus as A_{OLc} declines along its response curve, the corresponding frequency increases in direct proportion, indicating a constant gain–bandwidth product. At f_{p21}, $A_{OLc} = A_{min1}$, giving GBW $= A_{min1}f_{p21}$. At f_{cc}, $A_{OLc} = 1$, giving GBW $= f_{cc}$. The constant gain–bandwidth product of A_{OLc} equates the last two GBW expressions for $f_{cc} = A_{min1}f_{p21}$. Thus the differentiator-feedback compensation increases the maximum attainable bandwidth by a factor equal to A_{min1}.

A key frequency, noted on the curves, begins to define the phase compensation requirements. The continuation of the single-pole A_{OLc} response requires an alignment of the A_{CL2} zero at F_{z2} with the A_{OL1} pole at f_{p21}. Circuit analysis defines F_{z2}, quantifying the alignment requirement as $f_{z2} = 1/2\pi(R_3 + R_4)C = f_{p21}$. Solving for R_4 defines this element's design equation as $R_4 = 1/2\pi f_{p21}C - R_3$. However, this equation depends on two unknowns, resistance R_3 and capacitance C. Basic resistance level considerations common to all op amp feedback guide the otherwise arbitrary selection of R_3, removing one unknown.

The independent stability requirements of the A_2 feedback define the value of the second unknown, C. This amplifier's local feedback must satisfy its own stability criterion, as interrelated with that of the composite. Figure 5.13 displays the expanded set of response curves that combine the stability analyses for the two circuit feedbacks. While complex, this figure simply adds the pertinent response curves of A_2 to the curves just examined in Fig. 5.12. Here, a second frequency alignment in the response curves defines the value of C. Stability conditions of the local A_2 feedback require that the $1/\beta_2$ response pole occur no later than the f_{i2} intercept. Aligning the pole of $1/\beta_2$ with f_{i2} produces 45° of phase margin for this loop, as described in Sec. 5.2. This alignment requires that $1/2\pi R_3 C = f_{i2}$, and solving for C yields $C = 1/2\pi R_3 f_{i2}$. However, this result introduces a new unknown, f_{i2}.

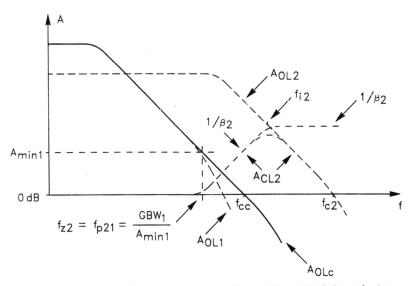

Figure 5.13 Adding the A_2 response curves to those of Fig. 5.12 defines the interrelated stability requirements of the composite and local feedbacks.

Geometric analysis of the response curves defines f_{i2} to yield the final design equation for C. In Fig. 5.13 a rising $1/\beta_2$ curve and a falling A_{OL2} curve define an isosceles triangle. Analysis shows the triangle's peak to be at the geometric mean of the frequencies defining the triangle's base. This makes $f_{i2} = \sqrt{f_{p21}f_{c2}}$, and combining this result with the previous expression for capacitance C produces the final design equation,

$$C = \frac{1}{2\pi R_3 \sqrt{f_{p21}f_{c2}}}$$

In a given application, the data sheets for the op amps selected define f_{c2} and generally f_{p21}. In some cases, circuit conditions create f_{p21}, requiring analysis or testing to determine this frequency. Given these two frequencies, choosing a value for R_3, based upon resistance level considerations, sets the value for C. Then, with known values for R_3 and C, the value for R_4 follows from the previously derived equation,

$$R_4 = \frac{1}{2\pi f_{p21}C} - R_3$$

These preceding design equations satisfy the stability requirements of both the local feedback of A_2 and the composite feedback of the overall circuit.

5.4.3   Selecting op amps for differentiator-feedback compensation

Simple application of the preceding design equations fails to provide the insight needed to select the two op amps at first. However, further examination of the response curves quantifies a relative speed requirement that guides the amplifier selection. This requirement relates the gain–bandwidth product GBW of A_2 to that of A_1. Successful implementation of the differentiator-feedback phase compensation requires that A_2 be a much faster op amp than A_1. The circuit's active zero depends on continued gain from A_{CL2} as the A_{OL1} gain rolls off. Fortunately, this requirement fits with composite amplifier capabilities due to a precision-versus-bandwidth compromise common to op amps. Selecting A_1 for precision assures good composite input characteristics but, generally, restricts this amplifier's bandwidth. A_2, as the second amplifier, does not require input precision, permitting optimization of bandwidth in this amplifier's choice.

Further examination of the response curves quantifies the speed difference requirement. Gain A_{CL2} counteracts the second pole of A_{OL1}, but A_{CL2} eventually introduces poles of its own. The A_{CL2} roll off at f_{i2}

introduces two coincident poles in the A_{OLc} response. One pole results from the interruption of A_{CL2}'s zero and the other results from the A_{OL2} roll off. To preserve the composite A_{OLc} response, these f_{i2} poles must be well above any $1/\beta_c$ intercept with this response. Otherwise, those poles introduce excessive phase shift at that critical intercept. Separating the composite amplifier's $1/\beta_c$ intercept and f_{i2} frequencies by a factor of three limits the added phase shift to around 45°. This combines with the 90° from A_{OLc}'s first pole, leaving a phase margin of $\phi_m = 180° - 135° = 45°$.

This factor of three translates to a relative op amp gain–bandwidth requirement in two steps. First, the case of $1/\beta_c = 1$ defines the most stringent requirement, and then $1/\beta_c < 1$ eases the requirement. For a feedback factor of unity, $\beta_c = 1$, $1/\beta_c = 1$, and the $1/\beta_c$ curve follows the unity gain or 0-dB axis of Fig. 5.13. There, it intercepts the A_{OLc} curve at this curve's crossover frequency f_{cc}. In this case, 45° or greater phase margin requires that $f_{i2} \geq 3f_{cc}$, where $f_{i2} = \sqrt{f_{p21}f_{c2}}$ and $f_{cc} = A_{min1}f_{p21}$ from before. Combining these three expressions and solving for f_{c2} defines the relative speed requirement of the two amplifiers for the $\beta_c = 1$ worst case. This solution produces the requirement $f_{c2} \geq 9(A_{min1})^2 f_{p21}$.

This intermediate result simplifies due to the constant gain–bandwidth product commonly found with op amps. The simplified result replaces the characteristic frequencies f_{c2} and f_{p21} of the amplifiers with their gain–bandwidth products. Most op amps produce a single-pole A_{OL} roll off over the majority of their useful frequency ranges. A 1:1 slope characterizes this single-pole roll off, making the product of gain and bandwidth a constant along this roll-off curve. In the example illustrated, A_{OL2} displays the most common op amp response, continuing its single-pole roll off all the way to the unity-gain crossover. There, at f_{c2}, $A_{OL2} = 1$, making $GBW_2 = f_{c2}$. Substituting this equality into the relative speed requirement from before produces $GBW_2 \geq 9(A_{min1})^2 f_{p21}$.

Next, conversion of f_{p21} to a gain–bandwidth product equivalent introduces an additional consideration. Gain A_{OL1} interrupts its single-pole roll off at f_{p21}, discontinuing the constant gain–bandwidth product for higher frequencies. Still, f_{p21} marks the last point on the curve's constant 1:1 slope and there $A_{OL1} = A_{min1}$. Thus $GBW_1 = A_{min1}f_{p21}$ for the single-pole region of A_{OL1}. Substituting this result into the relative speed requirement produces an insightful result of $GBW_2 \geq 9A_{min1}GBW_1$, where $GBW_2 = f_{c2}$ and $GBW_1 = A_{min1}f_{p21}$. Thus for $\beta_c = 1$, amplifier A_2 must have a gain–bandwidth product that exceeds that of A_1 by a factor of $9A_{min1}$.

Reduced feedback factors ease this requirement by a factor equal to β_c. Reducing β_c from unity moves the $1/\beta_c$ curve upward, away from the 0-dB axis. This movement pulls back the $1/\beta_c$ intercept with A_{OLc}, adding to the frequency separation between the $1/\beta_c$ intercept and the

f_{i2} poles. The 1:1 slope of the single-pole A_{OLc} response makes this movement a linear function of β_c. Thus reducing β_c from the unity case considered before moves the $1/\beta_c$ intercept from f_{cc} back to $\beta_c f_{cc}$. Then, the f_{i2} poles may move back by the same amount and retain the same 45° phase margin described before. This modifies the relative speed requirement for the two op amps to the final result

$$GBW_2 \geq 9\beta_c A_{min1} GBW_1$$

where $GBW_2 = f_{c2}$, $GBW_1 = A_{min1} f_{p21}$, and $\beta_c = R_1/(R_1 + R_2)$.

In the preceding, the multiplier $9\beta_c A_{min1}$ often imposes a demanding requirement upon the A_2 bandwidth. Fortunately, the nature of the differentiator-feedback compensation makes this requirement easier to meet. This compensation permits use of a lightly compensated amplifier for the A_2 function, extending this amplifier's gain–bandwidth product. In that case, the differentiator-feedback compensation produces a unity-gain stable composite formed with two op amps that lack such stability. The example described before assumed that A_1 lacked this stability, due to this amplifier's second pole. However, Fig. 5.13 shows a unity-gain stable A_{OL2} response for A_2.

Alternately, the A_{OL2} response can display a second pole between f_{i2} and f_{c2} without necessarily producing instability. Figure 5.14 shows the A_2 response curves for this case with a second A_{OL2} pole at f_{p22}, the frequency of the second pole of the second amplifier. Stability for the A_2 feedback loop depends on the A_{OL2} phase shift at the f_{i2} intercept. At f_{i2}, the f_{p22} pole does not significantly affect this phase shift as long as that pole occurs well above f_{i2}. Analysis shows that an $f_{p22} \geq 10f_{i2}$

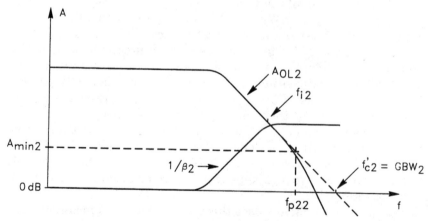

Figure 5.14 A second A_{OL2} pole disturbs the triangle of the Fig. 5.12 analysis, but a straight-line extension of A_{OL2} restores the analysis for extrapolation from previous results.

adds no more than 5.7° of phase shift at the f_{i2} intercept. Lightly compensated op amps that meet this f_{p22} requirement still suit the A_2 function in this application.

However, use of a lightly compensated A_2 requires modification of the previous design equation, $C = 1/2\pi R_3 \sqrt{f_{p21}f_{c2}}$. This equation depends on the frequency f_{c2}, which loses meaning in the lightly compensated A_2 case. The derivations of this capacitance design equation relied upon a geometric relationship in which f_{c2} defined one endpoint of a triangle base. Here, the second A_{OL2} pole interrupts the triangle's right side, negating the previous analysis. Still, a straight-line extension of A_{OL2}'s single-pole response restores the triangle, as shown by a dashed line in the figure. This dashed line defines a replacement for f_{c2} in the equation for C through a single-pole response extension that represents a constant gain–bandwidth product. At any point on this line, the gain coordinate times the frequency coordinate equals the GBW_2 of the original single-pole A_{OL2} response. Where this line extension crosses unity gain in Fig. 5.14, GBW_2 equals unity times f'_{c2}, making the right-hand endpoint of the new triangle's base $f'_{c2} = GBW_2$. Substituting GBW_2 for f_{c2} adapts the capacitance design equation for the case of a lightly compensated A_2 to produce

$$C = \frac{1}{2\pi R_3 \sqrt{f_{p21}GBW_2}}$$

5.4.4 Alternative differentiator-feedback compensation

The preceding phase compensation produces an active zero through differentiator feedback around a high-speed A_2. That compensation solution serves the case where precision dictates the choice of A_1 and speed determines the choice of A_2. The resulting combination permits use of a lightly compensated A_1 with A_2 compensating for A_1's second pole. In other cases, the roles reverse and A_2 produces the second pole, requiring compensation support from A_1. This condition often results from capacitance or high-current loading at the composite amplifier output, as supplied from A_2. Capacitance loading produces a second A_2 pole through a reaction with that amplifier's output impedance.[12] High-current loading produces a similar pole for power amplifiers, where the amplifier's output impedance increases with frequency, introducing increased loading effects at higher frequencies. For these cases, the two amplifiers of the composite structure switch functional roles. Then, differentiator feedback around A_1 produces the active zero, compensating for the second pole of A_2. Use of a high-speed A_1 in this alternative extends the composite bandwidth beyond that of A_2.

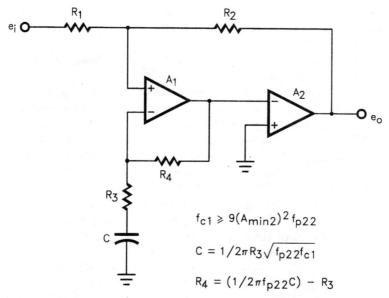

$$f_{c1} \geqslant 9(A_{min2})^2 f_{p22}$$

$$C = 1/2\pi R_3 \sqrt{f_{p22} f_{c1}}$$

$$R_4 = (1/2\pi f_{p22} C) - R_3$$

Figure 5.15 Reversing the amplifier roles of Fig. 5.11, differentiator feedback around A_1 compensates for the roll off of a second A_2 pole.

Figure 5.15 shows the alternative compensation with the three circuit changes required. Comparison of this circuit with the preceding one shows that the differentiator feedback moves to A_1, the input polarity assignments of the op amps reverse, and the overall composite connection switches to inverting. Moving the differentiator feedback to A_1 makes this amplifier produce the active zero. However, this move also requires reversing the op amp input polarities to prevent feedback interaction. Simply moving the differentiator feedback in Fig. 5.11 would connect both the local and the composite feedback networks to the inverting input of A_1. That connection places some of the feedback elements in parallel, making the circuit's two feedback networks interact. Then, each feedback network would influence both the open-loop and the closed-loop responses of the composite.

To avoid this interaction, Fig. 5.15 simply reverses the input polarity assignments of the two op amps. This change permits the separation of the two feedback connections through a new phase inversion in the composite feedback loop. The local differentiator feedback still connects to A_1's inverting input in a conventional negative feedback loop. However, the composite feedback now connects to A_1's noninverting input, relying on the phase inversion of A_2 to make the composite feedback negative. This separation of feedback connections allows the differentiator feedback to alter the composite's open-loop response with no interference from the composite feedback. Similarly, the composite feedback determines the circuit's closed-loop response with no interference from the differentiator feedback. However, this compensation alternative restricts the composite amplifier to the inverting configuration.

Moving the differentiator feedback to A_1 also increases the input errors of the composite amplifier. In previous composite connections, the high open-loop gain of A_1 isolates the composite input from the input errors of A_2 as they reflect to the composite input divided by a factor of A_{OL1}. However, the local feedback here reduces A_1's gain contribution to the composite loop, decreasing the reduction in A_2 input error effects from a factor of A_{OL1} to A_{CL1}. At lower frequencies, capacitor C blocks the A_1 feedback current, making $A_{CL1} = 1$. There, the composite loop drives A_1's noninverting input and the blocking action of C prevents current flow through R_3 and R_4. With no signal on R_4, A_1's output follows the feedback drive signal at this amplifier's noninverting input. Thus at lower frequencies, A_1 performs as a voltage follower, transmitting the feedback signal from R_1 and R_2 to the A_2 input. In the process, A_1 contributes only unity gain to the composite loop, and the full input errors of A_2 reflect to the composite input.

5.4.5   Optimizing the alternative differentiator compensation

At higher frequencies, capacitor C conducts feedback current, increasing A_{CL1} to produce the active zero. This increase reduces the effect of A_2 input errors and compensates for the second-pole roll off of that amplifier. Without this compensation, the second pole of A_{OL2} would compromise stability, as illustrated by the response curves of Fig. 5.16. These curves replicate those of Fig. 5.13 except for the subscripts designating the effects of the two amplifiers. Here, the two amplifiers switch roles, with A_1 producing the active zero to compensate for the second pole of A_2. In Fig. 5.16, the A_{OL2} response displays a two-pole roll off beginning at f_{p22}. This roll off would normally limit the use of A_2 to gains of A_{min2} or greater. However, the A_{CL1} response compensates for this roll-off pole with a zero at $f_{z1} = f_{p22}$. Linear summation of the A_{OL2} and A_{CL1} curves produces the composite open-loop response A_{OLc}. As before, the active zero compensation extends A_{OLc}'s single-pole response beyond the unity-gain crossover at f_{cc}. Above this frequency, A_{CL1} rolls off, adding two poles to the A_{OLc} response. One of these poles results from the A_{OL1} roll off and the other from the $1/\beta$ roll off at f_{i1}. The latter pole ensures stability for the A_1 local feedback loop.

Because these curves essentially replicate those of Fig. 5.13, conclusions drawn before extend to this case. Changing equation subscripts to account for the new amplifier roles adjusts these equations to the case considered here. This process defines the bandwidth improve-

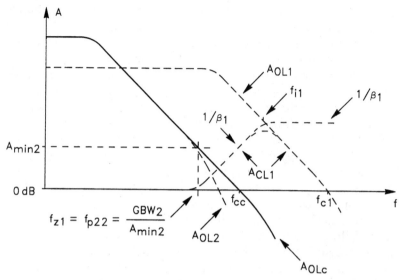

Figure 5.16 The differentiator feedback of Fig. 5.15 produces an A_{CL1} zero that compensates for the f_{p22} pole of A_{OL2}, extending the single-pole region of the composite A_{OLc}.

ment, component selection, and relative amplifier speed requirements that apply. For detailed discussions of the derivation of the equations, refer to earlier sections.

Extending the single-pole open-loop response with the composite amplifier increases the maximum attainable bandwidth, as summarized here, following the previous detailed discussion of Sec. 5.4.2. Using A_2 alone restricts the maximum attainable bandwidth to the frequency of the amplifier's second pole, f_{p22}. Adding A_1 to form the composite extends this bandwidth to the unity-gain crossover of the composite, f_{cc}. The constant gain–bandwidth product of the resulting A_{OLc} relates f_{cc} to f_{p22} through the relationship $f_{cc} = A_{min2}f_{p22}$. Thus the composite amplifier increases the maximum available bandwidth by a factor equal to A_{min2}, the gain level at which A_{OL2} encounters f_{p22}.

This bandwidth extension, and the circuit's stability, depend on the pole–zero cancellation produced by making $f_{z1} = f_{p22}$. Here, f_{z1} is the response zero added to the composite response by the differentiator feedback around A_1. This feedback must meet its own, local stability requirements by rolling off the zero's response at the f_{i1} intercept. Together, the composite and local feedback requirements define the design equations for the differentiator-feedback elements and quantified analysis of these requirements parallels the analysis in Sec. 5.4.2. Extrapolation of the earlier results to this case defines

$$C = \frac{1}{2\pi R_3 \sqrt{f_{p22} f_{c1}}}$$

For this calculation, the amplifier data sheet for A_1 defines f_{c1} and the data sheet for A_2 or empirical results define f_{p22}. Then, choosing a value for R_3 based on resistance-level considerations, sets the value for C. Given the selected values for R_3 and C, the value for R_4 follows from

$$R_4 = \frac{1}{2\pi f_{p22} C} - R_3$$

This feedback setting allows the A_1 high-frequency gain to counteract an A_2 accelerated roll off. However, A_1 must continue to supply gain to the composite loop well beyond the second A_2 pole. The poles of the A_{CL1} response must not occur at a frequency lower than $3f_{cc}$, where f_{cc} is the unity-gain crossover of the composite A_{OLc}. This imposes a relative bandwidth requirement for the two amplifiers paralleling that developed in Sec. 5.4.3. Translating the earlier result to this case produces the requirement

$$\text{GBW}_1 \geq 9\beta_c A_{\text{min2}} \text{GBW}_2$$

where $\text{GBW}_1 = f_{c1}$, $\text{GBW}_2 = A_{\text{min2}} f_{p22}$, and $\beta_c = R_1/(R_1 + R_2)$.

To meet this requirement more easily, lightly compensated op amps offer a greater gain–bandwidth product for the A_1 amplifier. It might seem that the lighter compensation would compromise stability, but A_1 does not have to be unity-gain stable to produce a composite structure. However, the quantity f_{c1}, required for the preceding equation for C, loses significance in this case. The derivations underlying that result depended on f_{c1} as the unity-gain crossover of a single-pole response. Replacing f_{c1} with GBW_1 resolves this issue, much as described in Sec. 5.4.3. Here, GBW_1 is the gain–bandwidth product of A_1 in this amplifier's region of single-pole response, and this represents the gain–bandwidth product reflected by data sheets for almost all op amps. As described in the earlier section, substituting GBW_1 for f_{c1} adapts the capacitor design equation for the case of a lightly compensated A_1, and then

$$C = \frac{1}{2\pi R_3 \sqrt{f_{p22} \text{GBW}_1}}$$

5.5   Feedback-Factor Phase Compensation

Most of the preceding compensation methods modify the A_{OLc} response, reducing the composite gain from its full $A_{OLc} = A_{OL1}A_{OL2}$ potential. There, local feedback around one of the two op amps sacrifices open-loop gain to shape the A_{OLc} response for a stable $1/\beta_c$ intercept. In this section, feedback-factor compensation realizes the composite's full open-loop gain potential by modifying only $1/\beta_c$, and the result improves both gain accuracy and settling time. However, the feedback-factor compensation only works for higher closed-loop gains and demands a more precise compensation selection. Response analysis again reduces that selection to a design equation.

5.5.1   Adding feedback-factor compensation

Figure 5.17 shows the composite amplifier with feedback-factor compensation. Just the feedback capacitor C_f stabilizes the composite loop. This capacitor adds a zero to the feedback factor, introducing phase lead to counteract phase lag. As frequency increases, the capacitor gradually increases the feedback factor to the maximum of unity. At unity, a feedback pole levels off the feedback factor with the capacitor providing short-circuit feedback. In equation form,

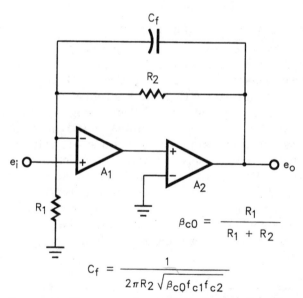

$$\beta_{c0} = \frac{R_1}{R_1 + R_2}$$

$$C_f = \frac{1}{2\pi R_2 \sqrt{\beta_{c0} f_{c1} f_{c2}}}$$

Figure 5.17 For greater loop gain and reduced settling time, phase compensating the composite feedback avoids the previous gain reductions and response singularities of local feedback.

$$\beta_c = \beta_{c0} \frac{1 + R_2Cs}{1+(R_1 \| R_2)Cs}$$

where $\beta_{c0} = R_1/(R_1 + R_2)$ represents the dc value of the feedback factor. With a pole following the zero, the compensating phase lead of β_c eventually returns to zero so the phase lead condition only spans a limited frequency range between the zero and the pole. Significantly separating the two singularities produces a frequency span sufficient for phase compensation. From the β_c equation, this separation requires that $R_2 \gg R_1 \| R_2$, and simplifying this expression produces $1 + R_2/R_1 \gg 1$. Thus feedback-factor phase compensation only serves higher-gain applications.

Response analysis summarizes the feedback action, guiding the phase compensation selection. Figure 5.18 supports this analysis and illustrates the loop gain benefit of the compensation. There, the zero and the pole of β_c described earlier produce the inverse effects for

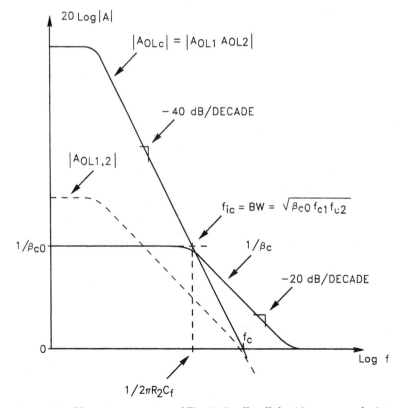

Figure 5.18   Phase compensation of Fig. 5.17 rolls off the $1/\beta_c$ curve, reducing the rate of closure at f_{ic}, but only for a restricted frequency range.

$1/\beta_c$. The $1/\beta_c$ curve rolls off with a single-pole response and then levels off at the unity-gain axis. Application of this response to phase compensation follows from the rate-of-closure criterion.[13] This criterion reflects stability conditions through the slope difference between A_{OLc} and $1/\beta_c$ at their intercept. A slope difference of 40 dB per decade corresponds to 180° of phase shift and oscillation. The two-pole uncompensated A_{OLc} response of the composite automatically produces a -40-dB per decade slope, threatening stability.

Most of the preceding phase compensation methods restore stability by reducing the A_{OLc} slope to around -20 dB per decade at the $1/\beta_c$ intercept. However, this sacrifices part of the loop gain potential of the uncompensated A_{OLc}. On closer examination, the rate-of-closure criterion only specifies the difference in slopes regardless of their individual magnitudes. So instead of reducing the A_{OLc} slope, feedback-factor compensation increases the $1/\beta_c$ slope. The result in Fig. 5.18 produces a region of -20-dB per decade slope for $1/\beta_c$. An appropriate positioning of the $1/\beta_c$ response places that slope at its intercept with A_{OLc}, making the slope difference a stable -20 dB per decade. This phase compensates the composite amplifier without sacrificing any of A_{OL}'s uncompensated gain or bandwidth.

This rate of closure suggests a 90° feedback phase shift at the intercept for a phase margin of 90°. However, both the pole and the zero of the $1/\beta_c$ response potentially reduce this margin. First, the illustrated $1/\beta_c$ pole occurs right at the f_{ic} intercept, reducing the phase margin to 45°, but this placement optimizes bandwidth while still preserving stability. In this compromise, the C_f bypass of R_2 rolls off the circuit's response to the signal, imposing a new bandwidth limit. Normally, f_{ic} defines the bandwidth limit as the point where the loop gain drops to zero but the addition of C_f introduces this second limit. Aligning the two bandwidth limits optimizes the net result. In addition, the zero of $1/\beta_c$ removes its response slope at some higher frequency and that also reduces the phase margin. Stability considerations require that this zero occur well above the $1/\beta_c$ intercept, which requires a significant frequency separation between pole and zero. That requirement typically restricts the feedback-factor compensation alternative to circuits having closed-loop gains of 30 or greater.

5.5.2 Optimizing the feedback-factor compensation

The selection of capacitor C_f produces the response singularity alignment described as guided by a design equation. From the figure, this alignment forces the equality $1/2\pi R_2 C_f = f_{ic}$, for which $C_f = 1/2\pi R_2 f_{ic}$. Once again, f_{ic} must be expressed mathematically to complete the compensation design equation. Examination of the A_{OLc} and $1/\beta_c$ curves yields this result. Frequency f_{ic} represents the uncompensated intercept of these two curves. Without compensation, the $1/\beta_c$ curve continues at its dc value of $1/\beta_{c0}$, as indicated by the dashed line that intersects A_{OLc} at f_{ic}. There, $1/\beta_{c0}$ and A_{OLc} occupy the same point on the graph, making $1/\beta_{c0} = A_{OLc}$ at f_{ic}. This intersection virtually always occurs in a region where the individual op amp responses follow single-pole roll offs. There, $A_{OL1} \approx f_{c1}/f$ and $A_{OL2} \approx f_{c2}/f$, making the composite gain $A_{OLc} = A_{OL1}A_{OL2} \approx f_{c1}f_{c2}/f^2$. Then, intercept f_{ic} occurs where $A_{OLc} = 1/\beta_{c0} \approx f_{c1}f_{c2}/f_{ic}^2$, and solving for f_{ic} yields

$$f_{ic} = \text{BW} = \sqrt{\beta_{c0}f_{c1}f_{c2}}$$

where $\beta_{c0} = R_1/(R_1 + R_2)$. Continuing the analysis defines C_f by setting $1/2\pi R_2 C_f = f_{ic} = \sqrt{\beta_{c0}f_{c1}f_{c2}}$ and solving yields

$$C_f = \frac{1}{2\pi R_2 \sqrt{\beta_{c0}f_{c1}f_{c2}}}$$

Feedback-factor phase compensation demands greater attention to compensation accuracy. Unusual in phase compensation, this method fails with too great a compensating capacitance. In most cases, increasing the compensation capacitance yields more damping and a more stable response. However, making C_f too large degrades stability in this case. Increasing C_f moves the $1/\beta_c$ curve to the left in Fig. 5.18, drawing the zero of the $1/\beta_c$ response closer to the intercept with A_{OLc}. As described, this $1/\beta_c$ variation compromises frequency stability. Only a certain range of C_f values provides circuit stability, emphasizing the importance of the C_f design equation in component selection.

To evaluate the bandwidth improvement expressed before, consider a composite amplifier formed with two identical op amps. Then, $f_{o1} = f_{c2}$, and the bandwidth becomes $\text{BW} = \sqrt{\beta_{c0}}\, f_{c1}$. For a single op amp, βf_{c1} expresses bandwidth and making $\beta = \beta_{c0}$ produces equivalent feedback conditions for the comparison. Dividing this into the composite bandwidth expression shows an improvement factor of $1/\sqrt{\beta_{c0}}$. Further examination compares this improvement with that of the earlier gain-limiting compensation. There, bandwidth improves by a factor of $1/\beta_2$ and the phase compensation sets $\beta_2 = \sqrt{\beta_c}$. This makes the earlier improvement factor $1/\sqrt{\beta_c}$, where β_c is equivalent to the β_{c0} here. Thus the bandwidth improvement of the feedback-factor compensation remains the same as that of the gain-limiting compensation. However, the feedback-factor compensation increases the composite open-loop gain dramatically.

Further, the feedback-factor compensation produces better settling time. The settling time depends on all response poles and zeros in the feedback loop, as each response singularity adds a time-dependent term to the transient response equation. Feedback reduces the effects of these terms through corrective loop gain. The feedback-factor compensation method first produces more loop gain and second restricts most singularities to a region of high loop gain. Figure 5.18 demonstrates these characteristics beginning with a smooth, continuous roll off for A_{OLc}. A double pole initiates the roll off, but these poles occur at a low frequency where the loop gain remains very high. No other singularity affects A_{OLc} until around the crossover f_c. At this high frequency, additional poles have very low time constants, making their settling time effects insignificant. The only feedback pole in a critical region remains that of $1/\beta_c$, and adjustment of this pole optimizes the compromise between rise and settling times.

In contrast, most of the previous compensation methods either limit the corrective loop gain or introduce additional response singularities. The gain-limiting and differentiator-feedback methods limit loop gain and the integrator-feedback method adds an open-loop pole–zero pair. This pair introduces lower frequency response terms, which settle slowly following a transient. Known as an integrating frequency doublet, this pole–zero combination became notorious for its poor settling time.[14]

References

1. J. Graeme, "Composite Amplifier Hikes Precision and Speed," *Electronic Design,* p. 30, June 24, 1993.
2. J. Graeme, *Optimizing Op Amp Performance,* McGraw-Hill, New York, 1997.
3. J. Graeme, "Phase Compensation Perks up Composite Amplifiers," *Electronic Design,* p. 64, August 19, 1993.
4. Graeme, op cit., 1997.
5. Graeme, op cit., 1997.
6. Graeme, op cit., 1997.
7. D. Lewis, "Compensation of Linear IC Test Loops," *Electronics test,* May 1979.
8. Graeme, op cit., 1997.
9. Graeme, op cit., 1997.
10. G. Tobey, J. Graeme, and L. Hueslman, *Operational Amplifiers; Design and Applications,* McGraw-Hill, New York, 1971.
11. Graeme, op cit., 1997.
12. Graeme, op cit., 1997.
13. Graeme, op cit., 1997.
14. J. Dostal, *Operational Amplifiers,* Butterworth-Heinmann, Stoneham, Maryland, 1993.

6

Differential-Output Amplifiers

The difference amplifier configuration of Sec. 1.3 provides common-mode rejection of noise signals coupled into an amplifier's input lines. Extending differential operation to the amplifier's output permits a similar rejection of noise signals coupled there. With differential-output lines, the circuitry following the amplifier can again employ common-mode rejection to remove coupled noise. For this purpose, the differential noninverting amplifier develops a differential output signal in response to a differential or single-ended input signal. In the process, this amplifier retains the common-mode rejection of input signals and largely removes the common-mode rejection ratio (CMMR) limitations of the preceding difference amplifier. In addition, the differential noninverting amplifier improves output voltage swing characteristics.

The basic differential noninverting amplifier still transfers any common-mode input signal directly to a common-mode output. This transfer does not necessarily interfere with the differential-output signal and the circuit's common-mode rejection continues to isolate the differential and common-mode output components. However, with long output connecting lines, the common-mode output signal produces a differential error signal through line impedance imbalances. For those cases, adding common-mode feedback to the basic configuration removes the output common-mode signal. Alternately, Chap. 7 presents a differential-output instrumentation amplifier that also removes this common-mode signal.

The differential noninverting amplifier also performs the task of single-ended-to-differential conversion. Most signals originate in the single-ended mode and, to benefit from common-mode rejection, they must be converted to the differential mode prior to line transmission. With or without common-mode feedback, the differential noninverting amplifier performs this conversion function directly but potentially suboptimizes either distortion or cost. To avoid these compromises, two feedback modifications adapt this circuit better to the conversion task. However, these modifications do impose a minimum circuit gain of around three. For applications requiring lower gain levels, an alternate single-ended-to-differential converter combines an inverting amplifier with a phase-equivalent noninverting amplifier. This alternative develops gains as low as unity without excessive phase distortion effects but does present a lower input impedance.

6.1   Differential Noninverting Amplifiers

The basic noninverting amplifier of Sec. 1.1.1 accepts a single-ended input signal and produces an amplified single-ended output signal. Simply combining two such amplifiers in a differential configuration accommodates differential input and output signals to fully exploit the noise reduction benefits of CMRR. Then, differential input signals benefit from the CMRR of the amplifier itself and differential output transmission permits a similar benefit in the circuitry that follows. For the input signal, the differential noninverting amplifier delivers extraordinary CMRR performance through an inherent cancellation of the two op amps' common-mode error signals. In addition, the two-amplifier combination doubles the output voltage swing and, in many cases, increases the slew rate. However, the bandwidth remains essentially the same as that of the basic noninverting amplifier.

6.1.1   Differential noninverting gain and CMRR

Connecting two noninverting op amp configurations with a common feedback network forms the differential connection shown in Fig. 6.1. Superposition analysis first displays the noninverting origin of this circuit. First, superposition grounding of the e_2 input produces a virtual ground at the inverting input of A_2. This essentially grounds the lower end of the R_G resistor, configuring A_1 as a noninverting amplifier to produce an output voltage of $e_{o1} = (1 + R_2/R_G)e_1$. An analogous superposition condition grounds the e_1 input and produces an A_2 output of $e_{o2} = (1 + R_2/R_G)e_2$. Thus the two op amps both perform as noninverting amplifiers but do so with a shared feedback resistor R_G.

This shared resistor connection develops the circuit's common-mode rejection by bootstrapping the input of one amplifier upon the input signal of the other. Continuing the superposition analysis demonstrates this rejection by considering the additional effect of a given input signal upon the opposite amplifier. For example, an e_2 signal transfers to the inverting input of A_2 where it also drives A_1 as an inverting amplifier. This condition produces an A_1 output component of $-(R_2/R_G)e_2$, and combining this component with that described earlier produces a net A_1

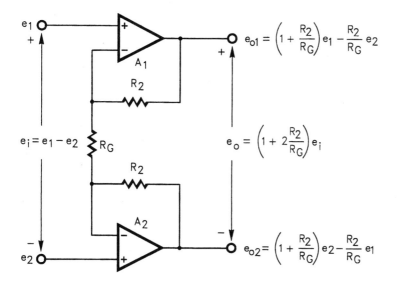

$$e_{o1} = \left(1 + \frac{R_2}{R_G}\right)e_1 - \frac{R_2}{R_G}e_2$$

$$e_o = \left(1 + 2\frac{R_2}{R_G}\right)e_i$$

$$e_{o2} = \left(1 + \frac{R_2}{R_G}\right)e_2 - \frac{R_2}{R_G}e_1$$

$$e_{ocm} = e_{icm} = (e_1 + e_2)/2$$

Figure 6.1   Combining two noninverting amplifiers with a common feedback network forms the differential noninverting amplifier to develop a differential output in response to a differential input.

output of $e_{o1} = (1 + R_2/R_G)e_1 - (R_2/R_G)e_2$. Analogously, the net A_2 output signal becomes $e_{o2} = (1 + R_2/R_G)e_2 - (R_2/R_G)e_1$. Together, the two op amps produce a differential output signal of

$$e_o = e_{o1} - e_{o2} = \left(1 + \frac{2R_2}{R_G}\right)(e_1 - e_2)$$

or

$$e_o = A_{Di}e_i$$

where $A_{Di} = 1 + 2R_2/R_G$ and $e_i = e_1 - e_2$. Here, A_{Di} represents the circuit's ideal voltage gain that the circuit sustains up to its response roll off. These response expressions transfer an amplified differential-input signal to a differential output while removing any signal common to e_1 and e_2 through the $(e_1 - e_2)$ subtraction.

Compared with the difference amplifier of Sec. 1.3, the differential noninverting amplifier also improves CMRR by removing much of the op amp error and resistor mismatch effects that imposed previous limits. Analysis defines the residual common-mode rejection limits using the model of Fig. 6.2. As before with the difference amplifier,

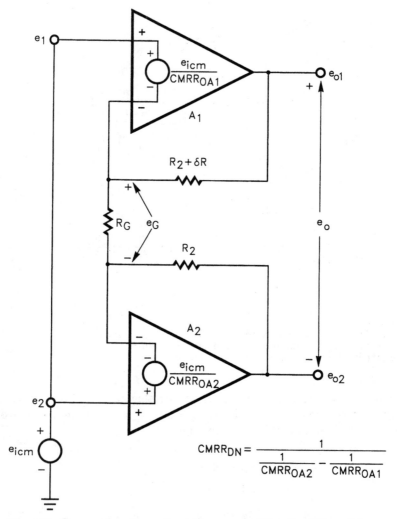

Figure 6.2 Compared to the earlier difference amplifier, the differential non-inverting amplifier improves common-mode rejection by removing common-mode signals from resistance imbalances and through cancellation of op amp common-mode rejection error signals.

the e_{icm}/CMRR_{OA} error sources between the op amp inputs model the inherent common-mode error effects of the op amps. Also repeated here, the δR_2 component of the upper R_2 resistor models the consolidated mismatch error of the circuit's resistors. That mismatch produces no CMRR limit for this case because the circuit isolates the resistors from the common-mode signal. Signal currents in these resistors result only from the voltage e_G developed across R_G. Initially neglecting the e_{icm}/CMRR_{OA} error sources to highlight the resistor effects makes $e_G = e_1 - e_2$, so only the differential input signal reaches R_G to create a feedback current for the circuit's two R_2 resistors. With no common-mode current flow in the resistors, resistor mismatches cannot develop error signals that would limit CMRR. Thus for first-order analysis, the δR component modeled only introduces a gain error to the circuit's differential response.

In practice, only the op amp e_{icm}/CMRR_{OA} error signals limit CMRR for the differential noninverting amplifier and then only through a residual error resulting from the difference between the two error signals. Analysis defines this residual through the definition of common-mode rejection, $\text{CMRR} = A_D/A_{CM}$, where A_D and A_{CM} represent the differential and common-mode gains of the circuit. From the preceding discussion, $A_{Di} = e_o/(e_1 - e_2) = 1 + 2R_2/R_G$ represents the differential gain over the circuit's useful frequency range. Analysis of Fig. 6.2 defines an expression for $A_{CM} = e_o/e_{icm}$ by first assuming the common-mode condition of $e_1 = e_2$. This reduces e_G to $e_G = e_{icm}/\text{CMRR}_{OA2} - e_{icm}/\text{CMRR}_{OA1}$. Previous analysis showed that the circuit amplifies e_G by a gain of A_{Di} to now produce $e_o = A_{Di}e_G = A_{Di}e_{icm}(1/\text{CMRR}_{OA2} - 1/\text{CMRR}_{OA1})$, making $A_{CM} = e_o/e_{icm} = A_{Di}(1/\text{CMRR}_{OA2} - 1/\text{CMRR}_{OA1})$. Applying this result to $\text{CMRR} = A_D/A_{CM}$ produces the final result for the differential noninverting amplifier,

$$\text{CMRR}_{DN} = \frac{1}{1/\text{CMRR}_{OA1} - 1/\text{CMRR}_{OA2}}$$

Using matched op amps for A_1 and A_2, such as those of a dual op amp, makes $\text{CMRR}_{OA2} \approx \text{CMRR}_{OA1}$ for dramatically improved common-mode rejection. Note that CMRR here represents the linear expression of the rejection and may require conversion from the logarithmic form often expressed as CMR in decibels. For this conversion, $\text{CMRR} = 10^{\text{CMR}/20}$.

6.1.2 Differential noninverting output swing

Compared with the earlier difference amplifier, the differential noninverting amplifier also improves the output voltage swing and potentially increases the slew rate. Voltage swing doubles due to the two amplifiers that now produce the circuit's output signal and gain. As with the difference amplifier, the output voltage swings of the circuit's op amps remain limited by their saturation voltage levels. However, unlike the difference amplifier, the two amplifiers here deliver a differential output from opposite polarity swings that double the normal limit. As a result, the peak-to-peak output voltage range of the differential noninverting amplifier routinely exceeds the net magnitude of the power supply voltages. For example, using a single op amp capable of ± 13-V output swing when powered from ± 15-V supplies produces a 26-$V_{p\text{-}p}$ output from a net 30-V power supply. However, the two op amps of the differential noninverting amplifier produce a ± 26-V or 52-$V_{p\text{-}p}$ output swing on the same 30-V supply.

At first, it would seem that the output swing doubling would also double the slew rate, but that also depends on specific signal conditions. To produce the differential output signal, one amplifier slews in one direction while the other slews in the opposite direction. This would double the differential slew rate except for the common-mode component of the circuit's output signal. As developed earlier, each circuit output supports a signal characterized by the response of the A_1 output, $e_{o1} = (1 + R_2/R_G)e_1 - (R_2/R_G)e_2 = e_1 + (R_2/R_G)(e_1 - e_2)$. There, both e_1 and the differential signal $e_1 - e_2$ place slewing demands upon the A_1 output, and this can preclude the expected slew rate doubling. Similarly, the A_2 output supports $e_{o2} = e_2 + (R_2/R_G)(e_2 - e_1)$, and together the two outputs support the common-mode signal $e_{ocm} = (e_{o1} + e_{o2})/2 = (e_1 + e_2)/2 = e_{icm}$.

The outputs of the individual op amps must slew in support of both the differential and the common-mode output signals e_o and e_{ocm}. Either signal can dominate slewing performance depending on the relative rates of change of the corresponding input signals. Ideally, the output of A_1 responds at a rate of $\delta e_{o1}/\delta t = \delta e_1/\delta t + (R_2/R_G)(\delta e_1/\delta t - \delta e_2/\delta t)$ and the independent $\delta e_1/\delta t$ term here supports both the differential and the common-mode output signals. An analogous expression describes the ideal A_2 response as $\delta e_{o2}/\delta t = \delta e_2/\delta t + (R_2/R_G)(\delta e_2/\delta t - \delta e_1/\delta t)$ and includes an independent $\delta e_2/\delta t$ term. Comparison of these expressions with the op amps' slew rate specifications determines whether or not the op amps limit output slewing for given signal conditions. As expressed, the source of such limiting depends on both $\delta e_1/\delta t$ and $\delta e_2/\delta t$ and their difference. In the common-mode signal, these rates of change directly influence the requirement $\delta e_{ocm}/\delta t = (\delta e_1/\delta t + \delta e_2/\delta t)/2$. Frequently, this signal results from small lower-frequency noise introduced by parasitic power-line coupling. In such cases, this signal induces no slew rate limit and the differential noninverting amplifier does double the slew rate as compared with the difference amplifier. Then, the differential output signal determines the slew rate limit imposed by the op amps through the requirement $\delta e_o/\delta t = (1 + 2R_2/R_G)(\delta e_1/\delta t - \delta e_2/\delta t)$. In other applications, $\delta e_1/\delta t$ and $\delta e_2/\delta t$ may greatly exceed $(1 + 2R_2/R_G)(\delta e_1/\delta t - \delta e_2/\delta t)$, and there the common-mode signal places the greater demand upon slew rate. Such can be the case when monitoring a small difference signal in the presence of a larger or higher-frequency common-mode signal.

Common-mode-induced rate limiting can override the transmission of the desired differential signal. In a phenomenon sometimes called transient intermodulation distortion, driving an amplifier into slew rate limiting with one signal component produces a blanking effect that prevents the transmission of a smaller accompanying component. That condition can occur with the differential noninverting amplifier when common-mode signals induce rate limiting even though the differential signal only requires a rate of change below the slew rate limit of the amplifiers.

6.1.3   Differential noninverting bandwidth

The output swing doubling also increases the circuit's voltage gain by a factor of two. Then, for a given gain level, that required of each op amp drops in half, which would seem to double the bandwidth for the differential noninverting amplifier. However, a more detailed analysis reveals that the presence of two amplifier error signals at the circuit input cancels this benefit. Examination of the fundamental bandwidth limit of op amp circuits reveals this cancellation. With the typical single-op-amp circuit only one op amp produces a gain error signal at the circuit input. As the op amp's open-loop gain rolls off with frequency, this error signal increases, reducing the signal produced at the output to produce the bandwidth limit. This single-op-amp case produces[1] $BW = \beta f_c$, where β and f_c represent the feedback factor and unity-gain crossover frequency of the op amp.

With the differential noninverting amplifier, two op amps contribute to the input error signal that defines the bandwidth limit and produce an equivalent result. For analysis of this effect, Fig. 6.3 models the gain error signals of the two op amps with e_o/A signal sources between the op amp inputs, where A represents the open-loop gain of the op amps. These error sources model the differential input signals required by the op amps to support their e_{o1} and e_{o2} output signals. Their error effects reduce the signals reaching the op amps' inverting inputs to $e_1 - e_{o1}/A$ and $e_2 - e_{o2}/A$, thereby reducing the signal developed on R_G to $e_G = e_1 - e_2 - e_{o1}/A + e_{o2}/A$. As gain A rolls off with frequency, the error terms increase to produce the circuit's bandwidth limit. Analysis of the circuit with these signal conditions defines a frequency-dependent voltage gain of

$$A_D(f) = \frac{A_{Di}}{1 + A_{Di}/A}$$

where $A_{Di} = 1 + 2R_2/R_G$ represents the circuit's ideal gain. At the frequencies of practical bandwidth limits, open-loop gain A follows a single-pole roll off toward the f_c crossover frequency, making $A = -jf_c/f$ for

$$A_D(f) = \frac{A_{Di}}{1 + j\,(A_{Di}f/f_c)}$$

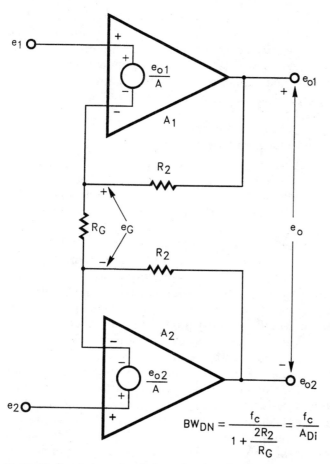

$$BW_{DN} = \frac{f_c}{1 + \dfrac{2R_2}{R_G}} = \frac{f_c}{A_{Di}}$$

Figure 6.3 Input error signals from the circuit's two op amps define the bandwidth limit of the differential noninverting amplifier.

The circuit's 3-dB bandwidth limit occurs when $A_D = 0.707A_{Di} = A_{Di}/\sqrt{2}$. That event occurs when the real and imaginary components of the denominator produce a magnitude of

$$\sqrt{(1)^2 + \left(\frac{A_{Di}f}{f_c}\right)^2} = \sqrt{2}$$

Then from the preceding equation, the condition $A_{Di}(f/f_c) = 1$ defines the frequency of the 3-dB bandwidth limit at $f = f_c/A_{Di}$ for

$$BW_{DN} = \frac{f_c}{A_{Di}}$$

where $A_{Di} = 1 + 2R_2/R_G$ and f_c represents the op amps' unity-gain crossover frequency. This BW_{DN} result parallels that of the basic noninverting amplifier, where $BW = f_c/A_{CL}$, as developed in Chap. 1.

6.2 Adding Common-Mode Feedback

The e_{ocm} signal described before potentially degrades the overall common-mode rejection of a system. The differential noninverting amplifier's two outputs support the common-mode signal $e_{ocm} = (e_{o1} + e_{o2})/2$, which transfers directly from the circuit input, making $e_{ocm} = (e_1 + e_2)/2 = e_{icm}$. As described before, circuitry following this amplifier can reject this e_{ocm} signal through its own common-mode rejection. Under ideal conditions, the two signal lines transmitting e_{ocm} to that circuitry maintain a separation of this signal from the desired differential signal. Then, common-mode rejection at the signal receiving end again removes the error effect of the common-mode component. However, impedance imbalances between the signal lines often convert a portion of the common-mode signal into a differential signal to potentially corrupt the signal transmission. Such unbalances commonly occur with very long lines and require removal of e_{ocm} prior to signal transmission, as provided by common-mode feedback.

6.2.1 Common-mode rejection of connecting lines

Before considering this feedback solution, Fig. 6.4 quantifies the problem of identifying those cases requiring the feedback. There, connecting lines transmit the e_o signal of the amplifier output through parasitic line impedances that alter the e'_o signal received at line end. In the figure, parasitic resistances R_P, parasitic capacitances C_P, and capacitance imbalance δC_P represent a simplified model of the line impedance condition. In practice, the lines introduce impedance imbalances other than the δC_P modeled through mismatches in resistive and inductive parasitics. However, the latter parasitics depend primarily on line length, and matched lengths generally make these imbalances secondary. In contrast, the parasitic capacitance from each line to ground depends on the many variables of the environment through which the lines are routed. There, even small differences in the capacitive results significantly degrade the precise impedance balance required to maintain high common-mode rejection for the system.

Analysis quantifies this degradation through the definition of the common-mode rejection ratio, $CMRR = A_D/A_{CM}$, where A_D and A_{CM} represent the differential and common-mode gains of a given transmission. Calculating these two gains for the line transmission defines

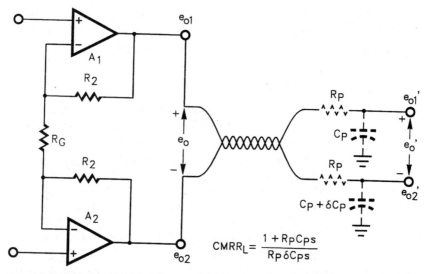

Figure 6.4 With the differential noninverting amplifier, line impedance imbalances degrade system common-mode rejection by converting a portion of the amplifier's common-mode output signal into a differential error, as characterized by a line common-mode rejection CMRR_L.

a line common-mode rejection CMRR_L. For the line, $A_{DL} = e'_o/e_o = (e'_{o1} - e'_{o2})/(e_{o1} - e_{o2})$ and assuming that $\delta C_P \ll C_P$ simplifies this portion of the analysis to yield $A_{DL} = 1/(1 + R_P C_P s)$. Next, including δC_P incorporates the impedance imbalance effect in $A_{CML} = e'_o/e_{ocm}$, where $e_{ocm} = (e_{o1} + e_{o2})/2$. This analysis yields $A_{CML} = R_P \delta C_P s/(1 + R_P C_P s)^2$. Then, $\text{CMRR}_L = A_{DL}/A_{CML}$ for this case produces

$$\text{CMRR}_L = \frac{1 + R_P C_P s}{R_P \delta C_P s}$$

This expression demonstrates a pole at zero frequency, due to δC_P, that rolls off CMRR_L, but a later zero terminates this roll off. At higher frequencies, this termination places an upper limit on the system common-mode rejection at $\text{CMRR}_L \leq C_P/\delta C_P$.

6.2.2 Common-mode feedback and common-mode range

Where the preceding line limit becomes significant, adding common-mode feedback removes the output common-mode signal to restore the differential output benefit. While this improves CMRR performance for the output lines, the common-mode feedback reintroduces resistor errors that potentially degrade the amplifier's own CMRR. In addition, this feedback potentially restricts the common-mode voltage

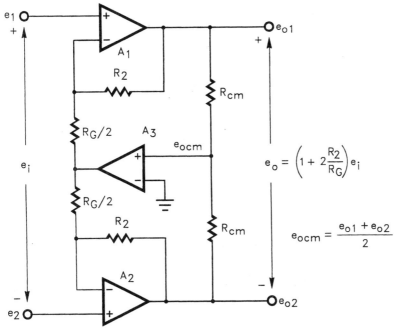

Figure 6.5 Adding common-mode feedback to the differential noninverting amplifier drives the output common-mode signal to essentially zero using the high gain of the A_3 feedback amplifier.

range of the input circuit. These compromises generally restrict the common-mode feedback option to applications with long connecting lines where impedance imbalances pose more serious limitations.

Figure 6.5 shows the common-mode feedback option with two R_{cm} sense resistors, feedback amplifier A_3, and a center tap formed in the R_G gain-set resistor. The R_{cm} resistors form a voltage divider to develop the output common-mode signal $e_{ocm} = (e_{o1} + e_{o2})/2$ at the resistor junction. Amplifier A_3 senses this signal and compares it to zero through the grounded connection of this amplifier's second input. Any deviation of this signal from zero causes A_3 to supply a correction voltage to the center tap formed by the two $R_G/2$ resistors. The resulting feedback signals developed by A_1 and A_2 remove the common-mode signal from the outputs of the differential noninverting amplifier. This modification only affects the circuit's common-mode response and the differential response still produces $e_o = (1 + 2R_2/R_G)e_i$. Note that the common-mode feedback requires connecting the R_{cm} sense point to the noninverting rather than the inverting input of A_3. A_3 drives A_1 and A_2 as inverting amplifiers, and the latter two amplifiers introduce the phase inversion required to establish negative rather than positive feedback. In addition, note that this connection places A_3 in a common feedback loop with both A_1 and A_2. As a general rule, any two amplifiers so connected require that one or the other produce a dominant lower-frequency pole to retain frequency stability. For most applications, this requires phase compensating A_3 or its feedback loop to develop the lower-frequency pole.

First consider the common-mode range restriction of this feedback. Setting $e_1 = e_2 = e_{icm}$ for the associated common-mode analysis of Fig. 6.5 results in the output signal expression

$$e_{o1} = e_{o2} = \left(1 + \frac{2R_2}{R_G}\right)e_{icm} - \frac{2R_2}{R_G}e_{o3}$$

Here both e_{o1} and e_{o2} depend on e_{o3}, the common-mode correction signal supplied by A_3. A_3 develops this signal in response to the common-mode output signal e_{ocm} and ideally forces this signal to zero. Then, the voltage divider formed by the R_{cm} resistors establishes the equal current condition $e_{o1}/R_{cm} = -e_{o2}/R_{cm}$. Substituting the e_{o1} and e_{o2} expressions in this last result and solving for e_{o3} produces

$$e_{o3} = \left(1 + \frac{R_G}{2R_2}\right)e_{icm}$$

This result reveals the potential new limit to the circuit's common-mode input range. As expressed, the output of A_3 must supply an e_{o3} greater than e_{icm}. Thus the output saturation limit of A_3 may occur for lower values of e_{icm} than would otherwise cause input saturation for A_1 and A_2. For the basic differential noninverting amplifier, that input saturation limit alone sets the common-mode input range. In this case, lower gain levels make $R_G \sim R_2$, demanding a significantly greater voltage range from the output of A_3. For example, a gain of $A_{Di} = 1 + 2R_2/R_G = 2$ requires that $R_G = 2R_2$ and $e_{o3} = 2e_{icm}$. In practical cases, this can reduce the common-mode input range by about a factor of two, and this compromise should be considered when adding common-mode feedback. The differential-output instrumentation amplifier presented in Chap. 7 avoids this common-mode range constraint.

6.2.3   Common-mode feedback and CMRR$_{DNd}$

Two common-mode rejection indicators characterize the operation of this modified differential noninverting amplifier. They reflect the rejection of input common-mode signals as seen from the differential and common-mode outputs of the circuit. The first, CMRR$_{DNd}$, follows from that described for the basic circuit and reflects the circuit's ability to isolate the signal developed differentially between the amplifier's two outputs from the effect of a common-mode input signal. The second, $CMRR_{DNc}$, reflects the circuit's ability to suppress the output common-mode signal resulting from a common-mode input signal. As described before, the basic differential noninverting amplifier does not suppress this latter signal and transmits the input common-mode signal directly to the circuit's two outputs with a gain of unity. As will be seen, adding common-mode feedback greatly reduces that gain to boost $CMRR_{DNc}$.

As always, the definition of common-mode rejection, $CMRR = |A_D/A_{CM}|$, guides the analyses of these common-mode rejection indicators, but with different output signals defining A_{CM} for the two cases. Figure 6.6 models the circuit of Fig. 6.5 for these analyses, showing the relevant amplifier and resistor error sources. There, the A_3 input error signal e_{o3}/A_{OL3} and the resistor matching errors δR_2 and δR_{cm} include the known or unexplored common-mode rejection circuit limitations. Previously, the common-mode rejection limit of the basic noninverting differential amplifier depended on the input error signals of A_1 and A_2. However, those error signals produce counteracting effects and their residual becomes negligible when compared to the A_3 and resistor error effects encountered here. As in the previous case, the δR_2 error modeled here consolidates the matching errors of all the R_2 and $R_G/2$ resistors. Absolute tolerance errors in these resistors introduce a gain error, but only their mismatches produce the common-mode rejection error. With the basic differential noninverting amplifier, the δR_2 error produced no common-mode rejection limit because no common-mode current flowed in the circuit's feedback resistors. However, the introduction of common-mode feedback here produces such a current and a resulting common-mode rejection error. The following analyses show that only δR_2 affects $CMRR_{DNd}$, and only δR_2 and A_{OL3} affect $CMRR_{DNc}$. δR_{cm} only produces secondary common-mode rejection effects through reactions with the error signals produced by δR_2 and A_{OL3}, making δR_{cm} negligible for first-order common-mode analyses.

Both $CMRR_{DNd}$ and $CMRR_{DNc}$ analyses of this case depend on the output signal components e_{o1} and e_{o2} that result from an input common-mode signal e_{icm}. Setting $e_1 = e_2 = e_{icm}$ for these analyses results in the output signal expressions

$$e_{o1} = \left(1 + \frac{2R_2}{R_G}\right)e_{icm} - \frac{2R_2}{R_G}e_{o3}$$

$$e_{o2} = \left[1 + \frac{2(R_2 + \delta R_2)}{R_G}\right]e_{icm} - \frac{2(R_2 + \delta R_2)}{R_G}e_{o3}$$

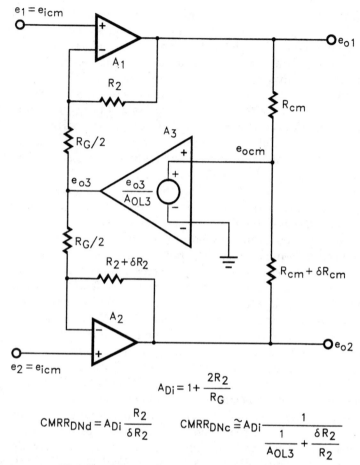

$$A_{Di} = 1 + \frac{2R_2}{R_G}$$

$$\text{CMRR}_{DNd} = A_{Di} \frac{R_2}{\delta R_2} \qquad \text{CMRR}_{DNc} \cong A_{Di} \frac{1}{\dfrac{1}{A_{OL3}} + \dfrac{\delta R_2}{R_2}}$$

Figure 6.6 Modeling Fig. 6.5 for common-mode rejection analysis represents the amplifier gain error and resistor mismatches that potentially set common-mode rejection limits.

From Sec. 6.2.2,

$$e_{o3} = \left(1 + \frac{R_G}{2R_2}\right) e_{icm}$$

and combining these results first defines

$$e_{o1} = 0$$

$$e_{o2} = -\frac{\delta R_2}{R_2} e_{icm}$$

As perhaps anticipated, the $e_{o1} = 0$ result reflects the fact that the resistor mismatches modeled in Fig. 6.6 do not directly affect the A_1 portion of the circuit. The model consolidates these mismatches in the A_2 portion and consequently produces a nonzero result for e_{o2}. While this represents a fictitious condition, the CMRR results that follow only depend on combinations of e_{o1} and e_{o2}, and those results reflect an accurate condition, with the analysis greatly simplified by this approach.

Given the expressions for e_{o1} and e_{o2}, first consider the determination of $CMRR_{DNd}$. As described before, this common-mode rejection indicator defines the differential output error signal resulting from a common-mode input signal. Common-mode gain A_{CMd} expresses this result as $A_{CMd} = (e_{o1} - e_{o2})/e_{icm}$ and substitution of the e_{o1} and e_{o2} results yields $A_{CMd} = \delta R_2/R_2$. Then, $CMRR_{DNd} = |A_{Di}/A_{CMd}|$ and $A_{Di} = 1 + 2R_2/R_G$ produce

$$CMRR_{DNd} = \left(1 + \frac{2R_2}{R_G}\right)\frac{R_2}{\delta R_2}$$

Note that this result predicts infinite CMRR for zero resistor mismatch, $\delta R_2 = 0$. In practice, resistor trimming increases CMRR to the background limit imposed by the A_1 and A_2 common-mode rejection errors neglected in this analysis. Note also that δR_{cm} does not appear in the $CMRR_{DNd}$ result, indicating that R_{cm} resistor mismatch does not affect the circuit's differential output.

6.2.4 Common-mode feedback and CMRR$_{DNc}$

Next, consider the common-mode rejection reflecting the suppression of the input-to-output transmission of an applied common-mode signal. Common-mode gain A_{CMc} expresses this transmission through $A_{CMc} = e_{ocm}/e_{icm} = (e_{o1} + e_{o2})/2e_{icm}$. Substituting the previous $e_{o1} = 0$ and $e_{o2} = -(\delta R_2/R_2)e_{icm}$ here defines $A_{CMc} = -\delta R_2/2R_2$, and $CMRR_{DNc} = |A_{Di}/A_{CMc}|$ yields

$$CMRR_{DNcR} = \left(1 + \frac{2R_2}{R_G}\right)\frac{R_2}{2\delta R_2}$$

This expression defines the $CMRR_{DNc}$ limit resulting from resistor mismatch only, and examination of the result reveals two key concepts. First, note that $CMRR_{DNcR} = CMRR_{DNd}/2$, showing that resistor mismatch affects the common-mode output signal twice as much as it does the differential output. Second, this new CMRR result also remains independent of δR_{cm}, so any R_{cm} resistor mismatch produces no primary effect upon the circuit's common-mode performance. Any such mismatch makes the circuit outputs deviate from the ideal in that $e_{o2} \neq -e_{o1}$. However, the $(e_{o1} - e_{o2})$ and $(e_{o1} + e_{o2})/2$ signal combinations that define the circuit's differential and common-mode output signals remain immune to this deviation.

One other factor does affect common-mode performance, the finite gain of A_3. Assumed infinite in the preceding analyses, the actual A_{OL3} gain produces a second limit to $CMRR_{DNc}$. To analyze just the effect of A_{OL3}, neglect the resistor mismatches considered earlier by setting δR_2 and δR_{cm} to zero. Then assuming zero input current for A_3, the current in the R_{cm} voltage divider establishes the condition

$$\frac{e_{o1} - e_{o3}/A_{OL3}}{R_{cm}} = \frac{e_{o3}/A_{OL3} - e_{o2}}{R_{cm}}$$

Rearranging this result produces $e_{o3}/A_{OL3} = (e_{o1} + e_{o2})/2$. By definition, $e_{ocm} = (e_{o1} + e_{o2})/2$, so $e_{ocm} = e_{o3}/A_{OL3}$, reflecting the gain error by which A_3 introduces a common-mode output signal and places a second limit upon $CMRR_{DNc}$. A_3 amplifies this error signal to produce the e_{o3} signal that ideally removes the common-mode signal from e_{o1} and e_{o2}. In the absence of a δR_2 error, the A_1 and A_2 noninverting amplifiers of the circuit both amplify the signal e_{o3} by the same gain and produce equal output results. Thus no differential output signal results from the A_{OL3} error, leaving $CMRR_{DNd}$ unchanged from that previously described for the δR_2 limit.

To quantify the new limit to $CMRR_{DNc}$, analysis again focuses on the common-mode response of the circuit through e_{icm} and e_{ocm}. Neglecting δR_2 and δR_{cm} for this case and assuming that $e_1 = e_2 = e_{icm}$, analysis of Fig. 6.6 defines $e_{o1} = e_{o2} = (1 + 2R_2/R_G)e_{icm} - 2(R_2/R_G)e_{o3}$ and $e_{o3} = A_{OL3}(e_{o1} + e_{o2})/2 = A_{OL3}e_{ocm}$. Combining these intermediate results produces $e_{ocm} = (e_{o1} + e_{o2})/2 = (1 + 2R_2/R_G)e_{icm}/(1 + 2R_2A_{OL3}/R_G)$ and $A_{CMc} = e_{ocm}/e_{icm} = (1 + 2R_2/R_G)/(1 + 2R_2A_{OL3}/R_G)$. Then, $CMRR = A_D/A_{CM}$, where $A_D = 1 + 2R_2/R_G$ yields the $CMRR_{DNc}$ limit imposed by A_{OL3},

$$CMRR_{DNcOA} = \left(1 + \frac{2R_2}{R_G}\right)A_{OL3}$$

This limit combines with that developed previously for resistor mismatch to define the net $CMRR_{DNc}$. As developed in Sec. 1.3.2, two such limits combine like paralleled resistances and

$$CMRR_{DNc} = \frac{1}{1/CMRR_{DNcR} + 1/CMRR_{DNcOA}}$$

Substitution of the previous $CMRR_{DNcR}$ and $CMRR_{DNcOA}$ results in this expression yields a complex result that offers little intuitive insight. However, considering the high-gain case produces an expression that reveals the relative importances of the amplifier and resistor errors. In that case, $A_D \gg 1$, and this requires that $2R_2/R_G \gg 1$ for

$$\mathrm{CMRR}_{\mathrm{DNc}} \cong A_D \, \frac{1}{1/A_{\mathrm{OL3}} + \delta R_2/2R_2}$$

where $A_D = 1 + 2R_2/R_G$. Typically, a very high value for A_{OL3} makes its associated error term negligible at lower frequencies and the δR_2 term dominates $\mathrm{CMRR}_{\mathrm{DNc}}$ there. At higher frequencies, the significances of the terms reverse due to the roll off of A_{OL3}. Thus to improve low-frequency $\mathrm{CMRR}_{\mathrm{DNc}}$, trim the R_2 resistors, and to improve high-frequency CMRR, choose a wide-band op amp for A_3.

6.3   Single-Ended-to-Differential Conversion

The common-mode rejection benefits of the preceding circuits depend on the transmission of a differential signal. However, most signals originate in single-ended rather than differential form. Converting a single-ended signal to differential mode prior to line transmission realizes the benefit of common-mode rejection noise reduction. Four circuits demonstrate this conversion beginning with the obvious solution and progressing to a balanced differential output structure having high input impedance.

6.3.1   Phase-balanced single-ended-to-differential conversion

The obvious circuit for single-ended-to-differential conversion combines a noninverting amplifier with an inverting equivalent, as shown in Fig. 6.7. There, amplifiers A_1 and A_2 develop output signals of equal magnitudes and opposite polarities, forming the differential output signal e_o. As a side benefit, these signal conditions produce second harmonic distortion cancellation in the differential signal. However, this circuit presents a low input impedance, equal to R_1, and a feedback factor difference for the two amplifiers introduces higher-frequency distortion. To produce the same gain magnitude, the noninverting and inverting amplifiers used here require different feedback networks, as shown and as discussed in Chap. 1.

The resulting feedback-factor difference introduces different bandwidth limits for the two amplifiers. For A_1, the feedback factor equals $\beta_1 = R_1/(R_1 + R_2)$, and for A_2, $\beta_2 = R_1/(2R_1 + R_2)$. From Chap. 1, the bandwidth of a single-op-amp circuit equals βf_c, where f_c is the unity-gain crossover frequency of the amplifier. With different values for β, the two op amps here would most likely impose different bandwidth limits. From a performance perspective, this would simply make the

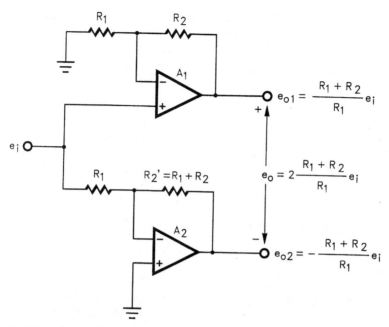

Figure 6.7 Straightforward approach for single-ended-to-differential conversion simply combines noninverting and inverting amplifiers having equal gain magnitudes, but differing feedback factors introduce phase error.

lower limit define the circuit's useful frequency range, except for associated phase errors. At frequencies much lower than that bandwidth limit, the differing phase shifts in the e_{o1} and e_{o2} output signals would likely produce a significant distortion in the differential output $e_o = e_{o1} - e_{o2}$.

Three circuit alternatives address the distortion and input impedance limitations of this straightforward implementation. To avoid the distortion error, the first alternative balances the feedback conditions of the two op amps through the addition of R_β in Fig. 6.8. That resistor connects between the inputs of A_1, where it remains isolated from the e_i input signal to avoid altering that amplifier's closed-loop gain. Then, only R_2 and R_G control the gain supplied by A_1. Making $R_G = R_1 R_2/(R_2 - R_1)$ as shown produces an A_1 gain of $1 + R_2/R_G = R_2/R_1$ to match the gain magnitude supplied by A_2 and produce the desired differential output signal. Here, the subtraction in the denominator of the R_G expression somewhat limits the practical range of that resistance but still permits circuit gains of unity and above.

While R_β does not affect the circuit's gain, it does alter the feedback factor of A_1 to balance the bandwidth and phase responses of the two op amps. For op amps, the feedback factor fundamentally equals the

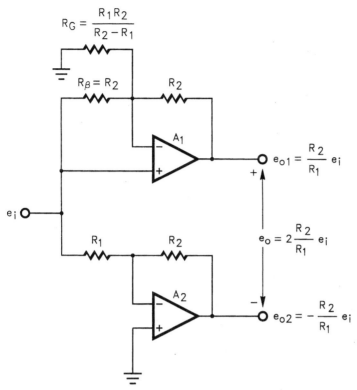

Figure 6.8 Adding the R_β feedback compensation resistor balances the feedback factors for the two op amps of Fig. 6.7 while retaining the equal gain magnitudes required for the differential output.

voltage divider ratio of the feedback network as seen from the amplifier's output.[2] As described in the referenced publication, this network includes R_2 and any feedback element that supports the gain error signal developed between the op amp's inputs. A_1 of the circuit impresses this signal upon both R_G and R_β, so both resistors influence the associated feedback factor. This makes the feedback factor for A_1 the voltage divider ratio $\beta_1 = (R_G \| R_\beta)/(R_G \| R_\beta + R_2)$. Setting $R_\beta = R_2$ reduces this to $\beta_1 = R_1/(R_1 + R_2)$ to match the feedback factor of A_2, equalize the op amp bandwidths, and avoid the phase distortion described earlier.

6.3.2 High-impedance single-ended-to-differential conversion

The preceding circuit removes the phase distortion encountered with the basic single-ended-to-differential conversion circuit. However, that solution does not address the low input impedance limitation and again makes the circuit's input impedance equal to the R_1 input

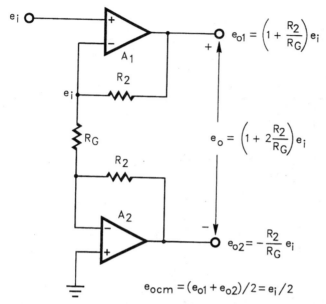

$$e_{o1} = \left(1 + \frac{R_2}{R_G}\right) e_i$$

$$e_o = \left(1 + 2\frac{R_2}{R_G}\right) e_i$$

$$e_{o2} = -\frac{R_2}{R_G} e_i$$

$$e_{ocm} = (e_{o1} + e_{o2})/2 = e_i/2$$

Figure 6.9 Using the differential noninverting amplifier for single-ended-to-differential conversion realizes high input impedance and retains equal feedback factors for the two op amps.

resistor of A_2. A second alternative employs the differential noninverting amplifier to establish the desired high input impedance. This circuit also retains the phase balance required to avoid distortion but, in its simplest form, introduces a common-mode output signal that can degrade CMRR performance. For gains of three or greater, feedback network adjustments remove this signal without disturbing the phase balance or input impedance.

Direct application of the differential noninverting amplifier to this conversion function produces the single-ended input configuration of Fig. 6.9. There, input signal e_i drives only the high impedance of an op amp input and the other circuit input returns to ground. From previous results, these input connections produce the desired single-ended-to-differential conversion as reflected in the differential output signal $e_o = (1 + 2R_2/R_G)e_i$. In addition, the balanced circuit structure produces identical feedback conditions for the two amplifiers, developing equal feedback factors and avoiding the previous phase distortion.

However, examination of the individual amplifier output signals reveals an accompanying common-mode signal component that could degrade long-line transmissions. This component results from the different circuit configurations presented to e_i by the two amplifiers of the circuit. Grounding the noninverting input of A_2 as shown transfers a virtual ground to that amplifier's inverting input. This makes A_1 appear as a noninverting amplifier to e_i and $e_{o1} = (1 + R_2/R_G)e_i$. Similarly, connecting the noninverting input of A_1 to e_i transfers e_i to that amplifier's inverting input. This makes A_2 appear as an inverting amplifier to e_i and $e_{o2} = -(R_2/R_G)e_i$. Together, the two output signals produce the common-mode output signal $e_{ocm} = (e_{o1} + e_{o2})/2 = e_i/2$. That signal potentially reacts with line impedance imbalances to produce a differential signal error as described in Sec. 6.2.1.

For longer line applications, common-mode feedback could be applied as described in Sec. 6.2. However, for the single-ended-to-differential conversion function a simpler alternative removes the common-mode output signal. Just changing one resistor and adding another balances the circuit to produce identical signal magnitudes at the two amplifier outputs and maintain identical feedback conditions. As shown in Fig. 6.10, these modifications alter the value of the lower R_2 resistor and add R_β. Increasing the lower R_2 resistor to $R'_2 = R_2 + R_G$ raises the inverting gain provided by A_2 to $-(1 + R_2/R_G)$ to match the gain magnitude provided by A_1. Then, the two amplifiers produce equal and opposite output signals, and $e_{ocm} = (e_{o1} + e_{o2})/2 = 0$. However, the modified R'_2 alters the feedback factor of A_2 and would reintroduce the higher-frequency distortion described at the beginning of

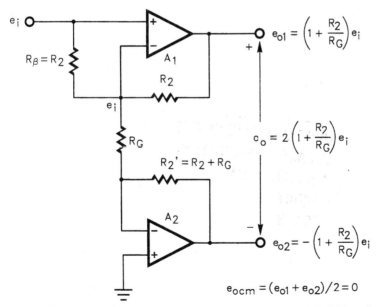

Figure 6.10 For single-ended-to-differential conversion, a modified R'_2 and an added R_β remove the output common-mode signal produced by the basic differential noninverting amplifier.

this section. To prevent this, adding R_β balances the A_1 feedback factor to match that of A_2. This balance simply requires making $R_\beta = R_2$, as described for the solution of Fig. 6.8.

This circuit alternative offers the best performance for single-ended-to-differential conversion in applications requiring voltage gains of around three or more. Below that range the circuit imposes impractical requirements on the feedback resistors. Analysis defines that gain limit through the differential output signal $e_o = e_{o1} - e_{o2}$, or

$$e_o = 2\left(1 + \frac{R_2}{R_G}\right)e_i$$

From this expression, making R_G infinite would deliver a gain of two. However, from Fig. 6.10 that resistor choice would open-circuit the signal connection between e_i and A_2 and would make R'_2 infinite. Both results would disable the circuit and practical applications of the circuit limit the maximum value of this resistance to around $R_G = 2R_2$ for a lower gain limit of three.

References

1. J. Graeme, *Optimizing Op Amp Performance*, McGraw-Hill, New York, 1997.
2. Graeme, op cit., 1997.

Instrumentation Amplifiers

Differential signal transmission and common-mode rejection of transmission errors characterize the role of the instrumentation amplifier. While sharing the differential-input nature of an op amp, the instrumentation amplifier includes committed feedback that retains high impedance at both amplifier inputs for superior immunity to line impedance loading. That immunity permits true differential sensing of a remote signal source for rejection of coupled noise and ground-potential difference errors. This chapter develops the gain, bandwidth, offset, and common-mode rejection responses of three instrumentation amplifier configurations. First, the three-op-amp configuration most commonly provides high overall performance, using two op amps in an input circuit that extracts the differential signal and a following difference op amp that removes the common-mode signal. Typically, the two input op amps dominate the bandwidth and offset performance and the third op amp dominates the common-mode rejection ratio (CMRR) result. Next, two op amps form a lesser used but simpler instrumentation amplifier with an application-specific benefit. Compared with the three-op-amp solution, this configuration restricts gain range, bandwidth, and signal swing but extends high-frequency common-mode rejection. Finally, adding a second difference amplifier to the three-op-amp configuration produces a differential-output instrumentation amplifier for the continuance of differential signal transmission. In addition, this solution offers greater output swing, slew rate, bandwidth, and common-mode rejection than the basic three-op-amp instrumentation amplifier.

7.1 Instrumentation Amplifier Advantages

Op amps and instrumentation amplifiers possess the same fundamental voltage-amplifier characteristics but in applications the instrumentation amplifier provides far superior rejection of coupled noise and ground-potential differences. Both amplifier types possess high-impedance differential inputs and a single-ended output, but they differ in their feedback conditions. Uncommitted feedback makes the op amp a versatile gain block capable of varied feedback control for the performance of many functions. However, the application of feedback to this amplifier compromises the impedance of one of its inputs. With the instrumentation amplifier, committed feedback generally restricts use to voltage gain functions, but this amplifier retains two high-impedance inputs for superior differential measurement results. To illustrate the significance of this difference, a comparison of voltage-amplifier monitoring of a remote source displays the performances of the two amplifier types.

In remote monitoring, the op amp produces the results demonstrated in Fig. 7.1. There, the long signal line connecting to the e_i source introduces a parasitic impedance Z_L. However, connecting the op amp as a noninverting voltage amplifier presents a high input impedance to the line and avoids the potential impedance-loading error. Still, this long line picks up the coupled noise signal e_n and the remote conditions introduce a ground-potential difference signal e_g. Those parasitic signals introduce error by altering the e'_i signal reaching the amplifier input, and $e'_i = e_i + e_g - e_n$ for $e_o = G(e_i + e_g - e_n)$, where $G = 1 + R_2/R_1$. Thus both the coupled noise signal e_n and the ground-potential difference signal e_g directly alter the e_o signal produced at the circuit output.

Under the same conditions, the instrumentation amplifier permits differential input sensing to greatly attenuate the effects of these error

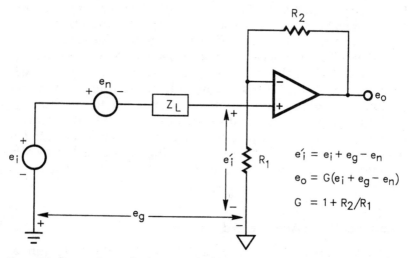

$$e'_i = e_i + e_g - e_n$$
$$e_o = G(e_i + e_g - e_n)$$
$$G = 1 + R_2/R_1$$

Figure 7.1 In remote monitoring, a noninverting op amp configuration avoids the loading error potentially introduced by line impedance Z_L but remains vulnerable to coupled noise e_n and ground potential difference e_g.

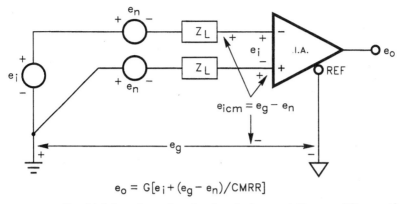

$$e_o = G[e_i + (e_g - e_n)/CMRR]$$

Figure 7.2 Two high-impedance inputs of an instrumentation amplifier avoid the line impedance effect as well as permitting differential input sensing to largely eliminate the effects of e_n and e_g.

signals. Figure 7.2 models the remote monitoring conditions for this solution, showing the same line and ground conditions as before. However, the instrumentation amplifier here adds a second input line for remote sensing of the signal common, producing true differential sensing of e_i. Having two high-impedance inputs, the instrumentation amplifier permits this two-line input connection without encountering loading errors from the Z_L line impedances. The op amp solution shown in Fig. 7.1 could also achieve differential sensing by returning the lower end of its R_1 resistor to the ground connection of the e_i source. However, that amplifier's low R_1 input impedance at this ground connection would introduce a loading error with a parasitic Z_L line impedance. Similarly, the difference amplifier connection of the op amp described in Sec. 1.3 would permit differential sensing of e_i, but again, low input impedances inherently produce line loading errors.

Having removed loading errors, the instrumentation amplifier next supplies common-mode rejection to remove the error signals induced by remote sensing. As shown in the figure, coupled noise e_n and ground-potential difference e_g again influence the signal coupled to the amplifier's inputs. However, the coupled noise produces equal e_n effects on the two input lines and signal e_i travels intact to the differential inputs of the instrumentation amplifier. These equal effects make e_n a common-mode signal at that amplifier's inputs. Similarly, e_g merely shifts both input lines by the same amount with respect to the common or reference connection of the instrumentation amplifier. Thus e_g also represents a common-mode signal at the instrumentation amplifier's inputs. The amplifier attenuates those errors by its CMRR prior to amplification, making $e_o = G[e_i + (e_g - e_n)/CMRR]$. Comparing this result with that of the previous op amp example shows an error signal reduction equal to the CMMR of the instrumentation amplifier.

7.2   Three-Op-Amp Instrumentation Amplifiers

While somewhat complex, this instrumentation amplifier configuration provides the most commonly used solution in differential signal monitoring. It first employs two op amps in a balanced input circuit that provides gain to the differential input signal. However, following this input circuit, the common-mode signal still remains, requiring a third op amp for its removal. Together, the three op amps produce an amplified single-ended output signal proportional to the differential component of the input signal. Typically, the two input op amps dominate the bandwidth and offset performance of the overall amplifier because they must provide gain. However, the third op amp must remove common-mode signals and it dominates common-mode rejection performance.

7.2.1   Response of the three-op-amp configuration

The three-op-amp instrumentation amplifier shown in Fig. 7.3, consists of the differential noninverting amplifier of Sec. 6.1 followed by the difference amplifier of Sec. 1.3. The example shown connects the difference amplifier for unity gain, as commonly done to permit an overall instrumentation amplifier gain as low as unity. Also, the A_1 and A_2 amplifiers depart from the conventional noninverting op amp connection in that the normally grounded feedback resistor R_G becomes a common feedback element for the two op amps. This departure creates the differential input character of the composite amplifier. Feedback control of A_1 and A_2 transfers the e_1 and e_2 input signals to the two ends of their common R_G gain-set resistor. This makes the voltage across R_G the differential input signal, or $e_i = e_1 - e_2$, and defines a common feedback current for A_1 and A_2.

The resulting current develops amplified voltages at the outputs of A_1 and A_2 and superposition analysis in Sec. 6.1 shows these signal voltages to be

$$e_{o1} = \left(1 + \frac{R_2}{R_G}\right)e_1 - \frac{R_2}{R_G}e_2$$

$$e_{o2} = \left(1 + \frac{R_2}{R_G}\right)e_2 - \frac{R_2}{R_G}e_1$$

The difference amplifier formed with A_3 subtracts these two signals to produce the final circuit output of

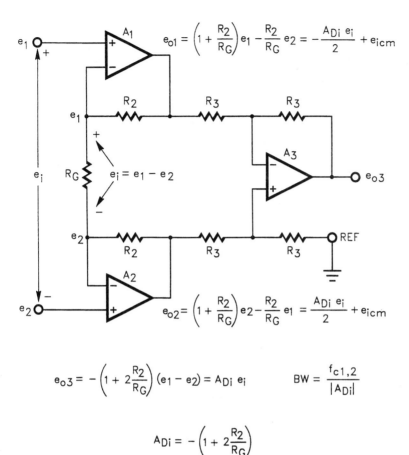

$$e_{o1} = \left(1 + \frac{R_2}{R_G}\right)e_1 - \frac{R_2}{R_G}e_2 = -\frac{A_{Di}\,e_i}{2} + e_{icm}$$

$$e_i = e_1 - e_2$$

$$e_{o2} = \left(1 + \frac{R_2}{R_G}\right)e_2 - \frac{R_2}{R_G}e_1 = \frac{A_{Di}\,e_i}{2} + e_{icm}$$

$$e_{o3} = -\left(1 + 2\frac{R_2}{R_G}\right)(e_1 - e_2) = A_{Di}\,e_i \qquad BW = \frac{f_{c1,2}}{|A_{Di}|}$$

$$A_{Di} = -\left(1 + 2\frac{R_2}{R_G}\right)$$

Figure 7.3 Three-op-amp instrumentation amplifier combines the differential noninverting amplifier with the difference amplifier to amplify the differential component of the input signals while removing their common-mode component.

$$e_{o3} = -\left(1 + 2\,\frac{R_2}{R_G}\right)(e_1 - e_2) = A_{Di}e_i$$

where $A_{Di} = -(1 + 2R_2/R_G)$ represents the circuit's ideal differential gain and $e_i = e_1 - e_2$ represents the differential input signal.

Further manipulation of these equations quantifies the source of an often surprising voltage swing limitation encountered with instrumentation amplifiers. These amplifiers must support both differential and common-mode signals internally and the two compete for the available voltage swing at the outputs of A_1 and A_2. Obviously, the amplified e_i differential signal introduces swing demands there. Less obviously, the common-mode signal shared by e_1 and e_2, $e_{icm} = (e_1 + e_2)/2$, introduces similar demands. As described, feedback transfers the e_1 and e_2 input signals to the two ends of the R_G resistor. Any signal common to e_1 and e_2 produces no current in R_G and thus receives no amplification from the circuit's R_2 feedback resistors. However, such a signal still transfers to the outputs of A_1 and A_2 with unity gain, producing a common-mode signal equal to e_{icm} at the outputs of these op amps. Rewriting the preceding expressions for the A_1 and A_2 output signals in terms of their differential and common-mode components provides a more direct evaluation of the associated output swing demands. In this process, recognizing that $e_i = e_1 - e_2$, $A_{Di} = -(1 + 2R_2/R_G)$, and $e_{icm} = (e_1 + e_2)/2$ produces

$$e_{o1} = -\frac{A_{Di}e_i}{2} + e_{icm}$$

$$e_{o2} = \frac{A_{Di}e_i}{2} + e_{icm}$$

Thus larger e_{icm} common-mode signals restrict the voltage swing range remaining to support the amplified differential signal $A_{Di}e_i$.

7.2.2 Bandwidth of the three-op-amp configuration

The bandwidth limit for this configuration generally occurs in the input amplifiers, due to the relative performance and gain demands of the three op amps. Selecting the A_1 and A_2 input amplifiers for precision performance reduces the amplified error effects of these op amps. However, precision and bandwidth place conflicting demands upon op amp architecture, and this choice typically restricts the bandwidth performance of the input op amps. Further, these two amps generate the circuit's voltage gain, introducing a second restriction through gain–bandwidth limitations. In contrast, op amp A_3 performs its difference-amplifier function following the gain provided by A_1 and A_2, and that gain removes the requirement for precision in this third op amp. Thus the A_3 architecture may be chosen to optimize bandwidth. Also, as a unity-gain difference amplifier, this op amp benefits from a larger feedback factor of $\beta_3 = 1/2$ and A_3 delivers a larger fraction of its maximum available bandwidth. That bandwidth becomes[1] BW $= \beta f_{c3} = f_{c3}/2$, where f_{c3} is the unity-gain crossover frequency of A_3. Then, selecting A_3 for $f_{c3} \gg 2f_{c1,2}$ assures that this amplifier does not introduce the dominant bandwidth limit, even when the gains of A_1 and A_2 reduce to unity.

In the typical case, identical op amps serve the A_1 and A_2 roles, making $f_{c1} = f_{c2} = f_{c1,2}$, and this suggests the bandwidth limit BW $= \beta f_{c1,2}$. However, as described in Sec. 6.1.3, the differential noninverting amplifier formed by A_1 and A_2 does not fulfill the requirement for the BW $= \beta f_c$ guideline of single-op-amp circuits. The bandwidth limit of the typical op amp circuit originates in the gain error signal developed between the differential inputs of the op amp. This error signal, e_o/A, represents the differential input signal required to support an e_o output signal through a finite open-loop gain A. At higher frequencies, the gain A rolls off, increasing the input error signal to reduce the output signal developed and produce the circuit's response roll off. For a single op amp circuit, this condition produces[2] BW $= \beta f_c$. However, the differential noninverting amplifier includes two op amps in its input circuit, requiring a return to the analysis of the input error signal effects upon bandwidth. As developed in Sec. 6.1.3, this analysis defines the bandwidth limit of the differential noninverting amplifier, and that limit transfers to the instrumentation amplifier considered here as

$$\text{BW} = \frac{f_{c1,2}}{|A_{Di}|}$$

where $A_{Di} = -(1 + 2R_2/R_G)$ and $f_{c3} \gg 2f_{c1,2}$.

7.2.3 Offset of the three-op-amp configuration

All three op amps of this configuration contribute to the circuit's output offset voltage, as modeled for analysis in Fig. 7.4. For this analysis, superposition grounding of the signal inputs focuses attention on the offset effects. As modeled, feedback controls the input circuit in support of the V_{OS1} and V_{OS2} input offset voltages of A_1 and A_2. This establishes equivalent voltages at the inverting inputs of these op amps and develops the voltage $V_{OS2} - V_{OS1}$ across the gain-set resistor R_G. In effect, this feedback state produces an input voltage condition equivalent to that of Fig. 7.3, for which $e_1 = -V_{OS1}$ and $e_2 = -V_{OS2}$. Making these substitutions in the e_{o1} and e_{o2} expressions of that earlier figure defines associated output offset components for A_1 and A_2 as $V_{OSO1} = -(1 + R_2/R_G)V_{OS1} + (R_2/R_G)V_{OS2}$ and $V_{OSO2} = -(1 + R_2/R_G)V_{OS2} + (R_2/R_G)V_{OS1}$.

Input bias currents I_{B1-} and I_{B2-} also develop output offset components for A_1 and A_2, as again controlled by feedback. Since feedback controls the voltage on R_G, it also controls the current in that resistor, preventing I_{B1-} and I_{B2-} from flowing there. Instead, feedback draws these currents through their associated R_2 feedback resistors. There, the currents develop the output offset voltage components $V_{OSO1} =$

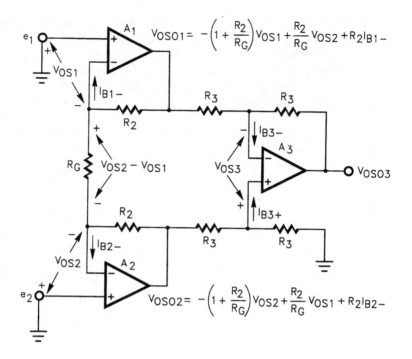

$$V_{OSO3} = \left(1 + 2\frac{R_2}{R_G}\right)(V_{OS1} - V_{OS2}) + R_2(I_{B2-} - I_{B1-}) - 2V_{OS3} + R_3(I_{B3-} - I_{B3+}$$

Figure 7.4 Individual offset voltages and input bias currents of this configuration's three op amps combine effects to produce the final output offset voltage V_{OSO3}.

$R_2 I_{B1-}$ and $V_{OSO2} = R_2 I_{B2-}$, which combine with those previously defined components to make the net output offsets of A_1 and A_2,

$$V_{OSO1} = -\left(1 + \frac{R_2}{R_G}\right)V_{OS1} + \frac{R_2}{R_G}V_{OS2} + R_2 I_{B1-}$$

$$V_{OSO2} = -\left(1 + \frac{R_2}{R_G}\right)V_{OS2} + \frac{R_2}{R_G}V_{OS1} + R_2 I_{B2-}$$

The difference amplifier formed with A_3 subtracts these V_{OSO1} and V_{OSO2} offsets to define the net output offset effects of A_1 and A_2. In this process, A_3 adds offset effects of its own through its input offset voltage and input bias currents. Due to this difference amplifier's feedback factor $\beta_3 = 1/2$, the $-V_{OS3}$ input offset voltage of A_3 receives a voltage gain of two, producing the output offset component $V_{OSO3} = -2V_{OS3}$. Similarly, I_{B3+} produces an A_3 input offset of $-(R_3/2)I_{B3+}$ that also receives a gain of two for the output component $V_{OSO3} = -R_3 I_{B3+}$. Finally, feedback controls the voltages of the A_3 input circuit to satisfy the conditions established by V_{OSO1}, V_{OSO2}, V_{OSO3}, and I_{B3+}. These multiple feedback constraints channel I_{B3-} away from the A_3 input circuit to the upper right R_3 resistor, producing the output offset component $V_{OSO3} = R_3 I_{B3-}$. Combining all these offset components produces the final A_3 output offset

$$V_{OSO3} = \left(1 + 2\frac{R_2}{R_G}\right)(V_{OS1} - V_{OS2}) + R_2(I_{B2-} - I_{B1-}) - 2V_{OS3}$$

$$+ R_3(I_{B3-} - I_{B3+})$$

At higher gain levels, the amplified offset voltages of A_1 and A_2 typically dominate, giving

$$V_{OSO3} \approx \left(1 + 2\frac{R_2}{R_G}\right)(V_{OS1} - V_{OS2})$$

7.2.4 CMRR of the three-op-amp configuration

All three op amps and their feedback resistors potentially affect the common-mode rejection of this instrumentation amplifier. However, in practical cases, just the difference-amplifier components dominate the final result. Figure 7.5 models the potential common-mode rejection errors with e_{id} input error signals for each op amp and a consolidated δR_3 resistance error for the A_3 difference amplifier. For each e_{id} error signal, $CMRR_{OA}$ represents the fundamental common-mode rejection ratio of a given op amp. The δR_3 error consolidates all of the resistor mismatches significant to the circuit's common-mode rejection ratio. Beginning the analysis with the input amplifiers, the full input common-mode signal e_{icm} exercises the inputs of A_1 and A_2, producing the corresponding input error signals $e_{id1} = e_{icm}/CMRR_{OA1}$ and $e_{id2} = e_{icm}/CMRR_{OA2}$. However, for matched input op amps, these error signals produce counteracting effects in a net signal of $e_{id2} - e_{id1} \approx 0$ across the R_G gain-set resistor. Any residual error resulting from mismatch between $CMRR_{OA1}$ and $CMRR_{OA2}$ typically remains insignificant in comparison to the other errors described later. Further, the very small residual signal voltage developed across R_G produces negligible feedback current for these amplifiers. As a result, any mismatch in the circuit's R_2 resistors produces no significant degradation in the common-mode rejection ratio.

While the input amplifiers produce little common-mode rejection error, A_1 and A_2 transfer the common-mode rejection task to the A_3

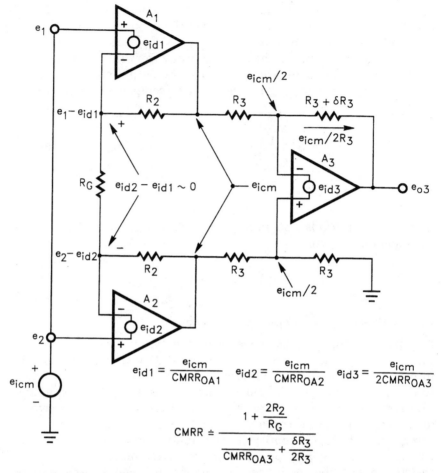

Figure 7.5 Three-op-amp e_{id} errors and a δR_3 resistance error potentially limit the CMRR of the three-op-amp instrumentation amplifier but error cancellation mitigates the error effects of the two input amplifiers.

difference amplifier. Because e_{icm} produces essentially no feedback currents for A_1 and A_2, these op amps act as voltage followers that signal and transfer e_{icm} to their two outputs. There, e_{icm} drives the inputs of the difference amplifier and produces an e_{id3} input error signal, with no counteracting error signal to remove its effect. Also, the transferred e_{icm} signal develops a feedback current that reacts with the resistance error δR_3. Together, the op amp and resistance errors of the difference amplifier dominate the common-mode rejection performance of the three-op-amp instrumentation amplifier. Analysis quantifies the result through the definition of the common-mode rejection ratio, $\text{CMRR} = |A_D/A_{CM}|$, where A_D and A_{CM} represent the circuit's differential and common-mode gains. From Sec. 7.2.1, the circuit produces a differential gain of $A_{Di} = -(1 + R_2/R_G)$ over the circuit's useful frequency range, defining the intermediate result, $\text{CMRR} = (1 + R_2/R_G)/|A_{CM}|$.

Evaluation of the e_{id3} and δR_3 error effects then defines A_{CM} through their associated output error signals. First consider the effect of e_{id3} as produced by the attenuated common-mode signal reaching the inputs of A_3. The lower R_3 resistors of the circuit deliver $e_{icm}/2$ to the noninverting input of A_3, and feedback replicates this signal at this op amp's inverting input, making $e_{icm}/2$ the relevant common-mode input signal. In response, A_3 produces the input error signal $e_{id3} = e_{icm}/2\text{CMRR}_{OA3}$. The upper R_3 resistors of the difference amplifier define a gain of two for $-e_{id3}$ to develop a component of output common-mode error equaling $e_o = -e_{icm}/\text{CMRR}_{OA3}$. Next, consider the effect of δR_3 in its reaction with the common-mode signal current of the difference amplifier. The e_{icm} signal at the inputs of the A_3 difference amplifier produces a feedback current equal to $e_{icm}/2R_3$ that flows through the δR_3 resistance error modeled. In practice, all four of the R_3 resistors contribute mismatch errors, but the modeled δR_3 simplifies the analysis and produces an equivalent result. This resistance error produces an output component of the common-mode error equal to $e_o = -e_{icm}\delta R_3/2R_3$.

Combining the e_{id3} and δR_3 error effects defines A_{CM} to complete the CMRR analysis for the three-op-amp configuration. Together, the two effects produce an output error signal of $e_o = -e_{icm}/\text{CMRR}_{OA3} - e_{icm}\delta R_3/2R_3$ for a common-mode gain equal to

$$A_{CM} = \frac{e_o}{e_{icm}} = -\frac{1}{\text{CMRR}_{OA3}} - \frac{\delta R_3}{2R_3}$$

where CMRR_{OA3} is the common-mode rejection ratio of op amp A_3 and $\delta R_3/R_3$ represents the net mismatch of the difference amplifier's two resistor networks. Combining this A_{CM} result with the definition of the common-mode rejection ratio, $\text{CMRR} = |A_D/A_{CM}|$, defines the common-mode rejection of the three-op-amp instrumentation amplifier as

$$\text{CMRR} = \frac{|A_{Di}|}{1/\text{CMRR}_{OA3} + \delta R_3/2R_3}$$

where $A_{Di} = -(1 + 2R_2/R_G)$ approximates the differential gain over the circuit's useful frequency range.

A review of this equation reveals two significant perspectives on the common-mode rejection of this instrumentation amplifier, as separated by frequency ranges. At lower frequencies, CMRR_{OA3} typically remains large, relegating the control of the common-mode rejection to the resistance mismatch δR_3. Fortuitously, the effect of this mismatch only occurs after the differential signal receives its gain through A_{Di}. Thus the more sensitive lower-level input signals, which require higher A_{Di} gains, benefit from higher common-mode rejection. However, at higher frequencies, CMRR_{OA3} inherently rolls off, making the fundamental common-mode rejection of op amp A_3 the determinant of the circuit's common-mode rejection there. As described later, the two-op-amp and differential-output instrumentation amplifiers remove this high-frequency limitation.

7.3   Two-Op-Amp Instrumentation Amplifiers

Two op amps form a lesser used but simpler instrumentation amplifier with an application-specific benefit. While simpler, this configuration imposes a minimum gain of around two, reduces bandwidth by a factor of two, and restricts signal swing by internally amplifying common-mode signals. However, focus upon these limitations often overlooks the primary advantage of this configuration. The two-op-amp instrumentation amplifier essentially removes the op amp error limitation of the instrumentation amplifier's primary function, common-mode rejection. At higher frequencies, this feature often makes the two-op-amp solution a better choice.

7.3.1   Response of the two-op-amp configuration

The two-op-amp instrumentation amplifier shown in Fig. 7.6, fundamentally consists of two series-connected noninverting amplifiers. There, subscripts a and b differentiate the fundamental feedback resistors of the two op amps for ease of discussion, but in practical cases $R_{1a} = R_{1b} = R_1$ and $R_{2a} = R_{2b} = R_2$. As before, this structure presents two high-impedance inputs to a signal source and develops a single-ended output signal. Adding a common feedback resistor R_G again permits control of the associated differential-to-single-ended gain. As with the three-op-amp case, feedback control of A_1 and A_2 transfers the e_1 and e_2 input signals to the two ends of the R_G resistor. This makes the voltage across R_G the differential input signal, or $e_i = e_1 - e_2$, and defines a common feedback current of $i_g = e_i/R_G$ for A_1 and A_2.

Superposition analysis quantifies the output signals of the two op amps by considering the effect of i_g separately. In the absence of i_g, op amp A_1 serves as a simple noninverting amplifier, amplifying e_1 to produce an intermediate signal component of $e_{o1} = (1 + R_1/R_2)e_1$. Introducing i_g produces a second component of the e_{o1} output signal as controlled by feedback conditions. The feedback of A_1 already controls the current in R_{2a} to make that resistor's voltage drop equal e_1, and

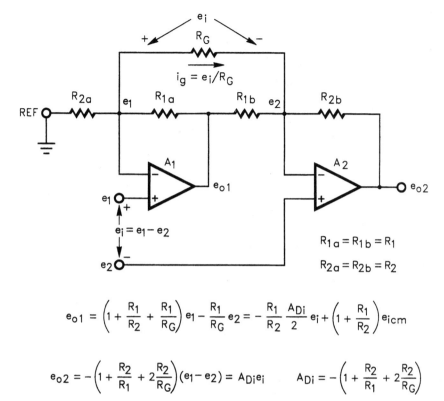

$$e_{o1} = \left(1 + \frac{R_1}{R_2} + \frac{R_1}{R_G}\right)e_1 - \frac{R_1}{R_G}e_2 = -\frac{R_1}{R_2}\frac{A_{Di}}{2}e_i + \left(1 + \frac{R_1}{R_2}\right)e_{icm}$$

$$e_{o2} = -\left(1 + \frac{R_2}{R_1} + 2\frac{R_2}{R_G}\right)(e_1 - e_2) = A_{Di}e_i \qquad A_{Di} = -\left(1 + \frac{R_2}{R_1} + 2\frac{R_2}{R_G}\right)$$

Figure 7.6 Two series-connected noninverting amplifiers with a common R_G feedback resistor form a simpler instrumentation amplifier having improved high-frequency CMRR.

the i_g current reaching A_1's inverting input cannot flow in that resistor. Instead, feedback draws the i_g current through the R_{1a} resistor, developing an intermediate signal component of $e_{o1} = (R_1/R_G)e_i = (R_1/R_G)(e_1 - e_2)$. Together, the two signal components described produce an A_1 output signal of

$$e_{o1} = \left(1 + \frac{R_1}{R_2} + \frac{R_1}{R_G}\right)e_1 - \frac{R_1}{R_G}e_2$$

where $R_{1a} = R_1$ and $R_{2a} = R_2$.

A similar superposition analysis defines the output signal of A_2, but from three rather than two input signal effects. For this op amp, the signals e_2, e_{o1}, and i_g each produce an output signal component. In the absence of e_{o1} and i_g, op amp A_2 serves as a simple noninverting amplifier to e_2, producing the output signal component $e_{o2} = (1 + R_2/R_1)e_2$. Similarly, in the absence of e_2 and i_g, op amp A_2 serves as a simple inverting amplifier amplifying e_{o1}, producing the output signal component $e_{o2} = -(R_2/R_1)e_{o1}$. Finally, introducing i_g produces the third component of this signal through a current flow controlled by other circuit conditions. This current cannot flow into R_{1b} because the output signal of A_1 and the feedback signal of A_2 already control the voltage across that resistor. To maintain this control, feedback draws i_g through the R_{2b} resistor, developing the output signal component $e_{o2} = (R_2/R_G)e_i = (R_2/R_G)(e_1 - e_2)$. Together, the three output components described here produce an A_2 output signal of

$$e_{o2} = -\left(1 + \frac{R_2}{R_1} + 2\frac{R_2}{R_G}\right)(e_1 - e_2) = A_{Di}e_i$$

where $R_{1a} = R_{1b} = R_1$, $R_{2a} = R_{2b} = R_2$, $A_{Di} = -(1 + R_2/R_1 + 2R_2/R_G)$, and $e_i = e_1 - e_2$. Here A_{Di} represents the ideal differential gain of the instrumentation amplifier, as produced up to the circuit's bandwidth roll off.

Further evaluation of these equations quantifies the range limitations of gain and voltage swing for the two-op-amp instrumentation amplifier. First, this circuit restricts the minimum practical voltage gain because as $R_G \to \infty$, the magnitude of gain A_{Di} only reduces to $1 + R_2/R_1$. This contrasts with the unity-gain minimum of the previous three-op-amp solution. In practice, the two-op-amp configuration typically imposes a minimum gain magnitude of around two, which occurs when $R_2 = R_1$.

Next, manipulation of the e_{o1} response expression explains a major voltage range restriction of the two-op-amp instrumentation amplifier. As explained in Sec. 7.2.1, instrumentation amplifiers must internally support both differential and common-mode signals. In this two-op-amp circuit, these signals compete for the available voltage swing at the output of A_1. Obviously, the differential e_i signal introduces a swing demand there. Less obviously, the common-mode signal shared by e_1 and e_2, $e_{icm} = (e_1 + e_2)/2$, introduces a similar demand. Rewriting the expression for the A_1 output signal in terms of its differential and common-mode components provides a more direct evaluation of the associated output swing demands. In this process, recognizing that $e_i = e_1 - e_2$, $A_{Di} = -(1 + R_2/R_1 + 2R_2/R_G)$, and $e_{icm} = (e_1 + e_2)/2$ produces

$$e_{o1} = -\frac{R_1}{R_2}\frac{A_{Di}}{2}e_i + \left(1 + \frac{R_1}{R_2}\right)e_{icm}$$

As described, making $R_2 = R_1$ optimizes the circuit's minimum gain range, and this makes $e_{o1} = -(A_D/2)e_i + 2e_{icm}$. In comparison, the previous three-op-amp instrumentation amplifier produced an internal signal voltage of $e_{o1} = -(A_D/2)e_i + e_{icm}$. Thus both configurations introduce the same swing demand from the differential signal e_i, but the two-op-amp configuration introduces twice the demand from the common-mode signal e_{icm}. Hence, an amplified common-mode signal produces a more restricted voltage swing limitation for the two-op-amp configuration.

7.3.2 Bandwidth of the two-op-amp configuration

The bandwidth limit for this instrumentation amplifier can result from either or both of the circuit's two op amps, depending on their relative feedback factors. Their unity-gain crossover frequencies could also introduce a variable. However, in the typical case, identical op amps serve the A_1 and A_2 roles to provide matching input characteristics. This op amp choice also makes $f_{c1} = f_{c2} = f_{c1,2}$ to remove the crossover-frequency variable. Once again, the universal op amp guideline[3] BW = βf_c aids in the determination of the actual bandwidth with a potential modification introduced by two coincident response roll offs. Both op amps impose BW = βf_c limits, and the lower of the two can produce a dominant limit. As will be seen, making the two limits coincident optimizes the net result but requires a modified bandwidth guideline.

To determine the actual bandwidth, superposition analysis first defines the feedback factor for each op amp by grounding the signal input of the other. To determine β_1 of A_1, grounding the e_2 input makes the inverting input of A_2 a virtual ground at the right end of the R_G resistor. This effectively places R_G in parallel with R_{2a}, making that op amp's feedback voltage divider a combination of R_{1a} and $R_G \| R_{2a}$. From the associated divider ratio

$$\beta_1 = \frac{R_2}{R_1 + R_2 + R_1 R_2 / R_G}$$

where $R_{1a} = R_1$ and $R_{2a} = R_2$. Note that as $R_G \to \infty$, β_1 reverts to the simple $\beta_1 = R_2/(R_1 + R_2)$ of the basic noninverting amplifier formed with A_1.

Superposition grounding of the e_1 input for the analysis of β_2 produces an analogous result. Then, the inverting input of A_1 presents a virtual ground to the left end of R_G. Similarly, the low-impedance output of A_1 presents a virtual ground at the left end of R_{1b}. This combination effectively places R_G and R_{1b} in parallel to form an A_2 feedback network composed of R_{2b} and $R_G \| R_{1b}$. With this combination,

$$\beta_2 = \frac{R_1}{R_1 + R_2 + R_1 R_2 / R_G}$$

where $R_{1b} = R_1$ and $R_{2b} = R_2$.

Either β_1, β_2, or their combination can determine the final bandwidth result, depending on the relative resistances of R_1 and R_2. Comparison of the expressions for β_1 and β_2 shows that they differ only in their numerators, as determined by R_1 and R_2. Thus a significant resistance difference between R_1 and R_2 will make one of the feedback factors the dominant bandwidth limit. For example, if $R_1 \gg R_2$, $\beta_1 \ll \beta_2$, making β_1 the dominant limit and BW $= \beta_1 f_{c1,2}$. Under these conditions, increasing R_2 increases β_1 and improves the bandwidth up to the point where $R_2 = R_1$. Beyond that point, making $R_2 \gg R_1$ switches the bandwidth limit to β_2 and then BW $= \beta_2 f_{c1,2}$. The optimum bandwidth occurs for $R_1 = R_2$, where $\beta_1 = \beta_2 = \beta_{1,2}$, and the two op amps impose coincident bandwidth limits. Figure 7.7 illustrates this optimization where making $R_1 = R_2 = R_{1,2}$ moves the f_{p1} and f_{p2} poles of the individual op amps together at $f_{p1,2}$. For this equal resistance case,

$$\beta_1 = \beta_2 = \frac{1}{2 + R_{1,2}/R_G}$$

Note that as $R_G \to \infty$, β_1 and β_2 revert to the simple $\beta = 1/2$ of the basic noninverting amplifier formed with equal feedback resistors. As described in Sec. 7.3.1, this equal resistance choice also optimizes the circuit's minimum gain.

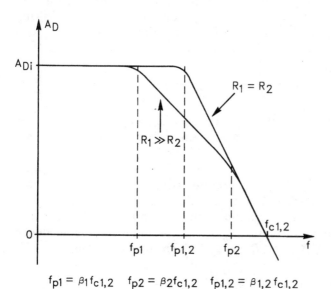

$$f_{p1} = \beta_1 f_{c1,2} \qquad f_{p2} = \beta_2 f_{c1,2} \qquad f_{p1,2} = \beta_{1,2} f_{c1,2}$$

Figure 7.7 For the two-op-amp instrumentation amplifier, equalizing the circuit's R_1 and R_2 feedback resistors moves its two poles together to optimize bandwidth.

To determine the final bandwidth for this $R_1 = R_2$ case, the BW = βf_c guideline requires an adjustment, as also illustrated by Fig. 7.7. There, two response curves represent the overall circuit's A_D gain for the cases where $R_1 \gg R_2$ and $R_1 = R_2$. For the first case, the f_{p1} and f_{p2} poles represent the individual bandwidth limits of the circuit's two op amps as expressed by $f_{p1} = BW_1 = \beta_1 f_{c1}$ and $f_{p2} = BW_2 = \beta_2 f_{c2}$. Under these conditions, the BW = βf_c guideline holds for the individual op amps, and it continues to hold for the instrumentation amplifier as long as these two poles remain well separated in frequency. That guideline depends on a response roll off initiated by a single pole,[4] and the well-separated f_{p1} and f_{p2} poles of the $R_1 \gg R_2$ case illustrated retain this condition for the overall amplifier. For this case, BW = f_{p1} = $\beta_1 f_{c1}$ and, while not illustrated, an $R_1 \ll R_2$ condition produces an analogous result of BW = $f_{p2} = \beta_2 f_{c2}$.

7.3.3 Optimum bandwidth of the two-op-amp configuration

For the optimal $R_1 = R_2$ case, the two $f_{p1,2}$ poles still represent the individual bandwidth limits of the circuit's two op amps, and the BW = βf_c guideline still holds for the individual amplifiers. However, as described in Secs. 6.1.3 and 7.2.1, a multiple-op-amp input circuit potentially does not fulfill the requirement for the BW = βf_c guideline of single-op-amp circuits. The bandwidth limit of the typical op amp circuit originates in the gain error signal developed between the differential inputs of the op amp and produces the BW = βf_c result. This guideline does not generally remain valid for the instrumentation amplifier because the two $f_{p1,2}$ poles initiate the response roll off with a double pole. In this case, both op amps contribute to the circuit's response roll off through their input error signals. For the cases where $R_1 \gg R_2$ or $R_2 \gg R_1$, this instrumentation amplifier produces $\beta_2 \gg \beta_1$ or $\beta_1 \gg \beta_2$, and only one op amp determines the bandwidth result through that amplifier's dominant input error signal and BW = βf_c still applies.

However, the $R_1 = R_2$ optimum condition makes $\beta_1 = \beta_2$, and the input error signals of the two op amps contribute equally to the bandwidth limit of the two-op-amp instrumentation amplifier. This departure requires a return to fundamental bandwidth analysis as supported by Fig. 7.8, where $R_{1a} = R_{1b} = R_{2a} = R_{2b} = R$. There, e_o/A error sources between the op amp inputs model the input gain errors, where A represents the open-loop gain of the op amps. These error sources represent the differential input signals required by the op amps to support their e_o output signals. Their error effects reduce the signals reaching the op amps' inverting inputs to $e_1 - e_{o1}/A$ and $e_2 - e_{o2}/A$ and reduce the signal developed on R_G to $e_G = e_1 - e_2 - e_{o1}/A + e_{o2}/A$. As

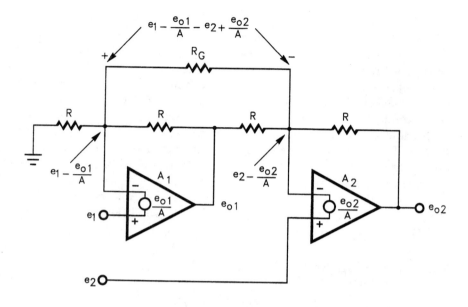

$$BW \approx \frac{f_{c1,2}}{2\left(2 + \dfrac{R}{R_G}\right)} = \frac{f_{c1,2}}{2|A_{Di}|}$$

Figure 7.8 With equal-feedback resistances, $R_1 = R_2 = R$, the two-op-amp instrumentation amplifier produces its optimum bandwidth, as defined by considering the e_o/A input error signals of the circuit's two op amps.

gain A rolls off with frequency, the error terms increase to produce the circuit's response roll off. Analysis of the circuit with these signal conditions defines a frequency-dependent voltage gain approximated by

$$A_D(f) \approx \frac{A_{Di}}{1 + 2A_{Di}/A}$$

where $A_{Di} = 2 + 2R_2/R_G$ represents the circuit's ideal differential gain. This approximation represents applications with higher levels of A_{Di} accurately, but it adequately approximates lower levels as well.

At the frequencies of practical bandwidth limits, open-loop gain A follows a single-pole roll off toward the f_c crossover frequency, making $A = -jf_c/f$ for

$$A_D(f) \approx \frac{A_{Di}}{1 + 2j\,A_{Di}f/f_c}$$

The circuit's 3-dB bandwidth limit occurs when $|A_D| = 0.707\,|A_{Di}| =$

$|A_{Di}|/\sqrt{2}$. That event occurs when the real and imaginary components of this denominator produce a magnitude of

$$\sqrt{(1)^2 + \left(2\,\frac{A_{Di}f}{f_c}\right)^2} = \sqrt{2}$$

Then, the condition $|2A_{Di}f/f_c| = 1$ defines the frequency f of the 3-dB bandwidth limit as $f = f_c/2|A_{Di}|$, and for the two-op-amp instrumentation amplifier having $R_1 = R_2$,

$$\text{BW} \approx \frac{f_{c1,2}}{2\,|A_{Di}|}$$

where $|A_{Di}| = 2 + 2R_2/R_G$ and f_c represents the op amps' unity-gain crossover frequency.

7.3.4   Offset of the two-op-amp configuration

The offset errors of the circuit's two op amps combine to produce a net output offset voltage, as modeled for superposition analysis in Fig. 7.9. There, grounded signal inputs focus the analysis on the effects of offset errors. In this condition, feedback controls the input circuit in support of the V_{OS1} and V_{OS2} input offset voltages of A_1 and A_2. This establishes these voltages at the inverting inputs of the respective op amps and develops the voltage $V_{OS2} - V_{OS1}$ across the gain-set resistor R_G. In effect, this feedback state produces an input voltage condition equivalent to that of Fig. 7.6, for which $e_1 = -V_{OS1}$ and $e_2 = -V_{OS2}$. Making these substitutions in the e_{o1} and e_{o2} expressions of that earlier figure defines the associated output offset components for A_1 and A_2 as $V_{OSO1} = -(1 + R_1/R_2 + R_1/R_G)V_{OS1} + (R_1/R_G)V_{OS2}$ and $V_{OSO2} = (1 + R_2/R_1 + 2R_2/R_G)(V_{OS1} - V_{OS2})$.

Input bias currents I_{B1-} and I_{B2-} also develop output offset components for A_1 and A_2, as again controlled by feedback. Since feedback controls the voltage on R_G, it also controls the current in that resistor, preventing I_{B1-} and I_{B2-} from flowing there. For op amp A_1, feedback also sets the voltage on R_{2a} at V_{OS1} to similarly define the current in that resistor, and this prevents the flow of I_{B1-} there. Instead, feedback draws this current through R_{1a}, where it develops an output offset voltage component of $V_{OSO1} = R_{1a}I_{B1-}$. This combines with the previously defined component to produce a net A_1 output offset of

$$V_{OSO1} = -\left(1 + \frac{R_1}{R_2} + \frac{R_1}{R_G}\right)V_{OS1} + \frac{R_1}{R_G}\,V_{OS2} + R_1 I_{B1-}$$

where $R_{1a} = R_1$ and $R_{2a} = R_2$.

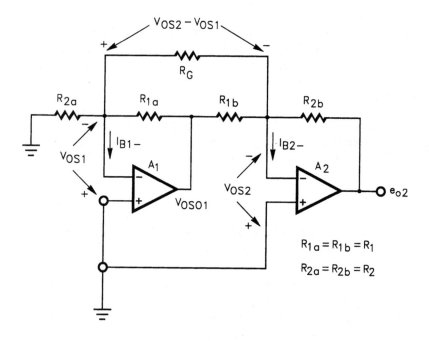

$$V_{OSO1} = -\left(1 + \frac{R_1}{R_2} + \frac{R_1}{R_G}\right)V_{OS1} + \frac{R_1}{R_G}V_{OS2} + R_1 I_{B-}$$

$$V_{OSO2} = A_{Di}(V_{OS2} - V_{OS1}) + (I_{B2-} - I_{B1-})R_2 \qquad A_{Di} = -\left(1 + \frac{R_2}{R_1} + 2\frac{R_2}{R_G}\right)$$

Figure 7.9 Grounded signal inputs focus signal analysis on the output offset produced by the circuit's two op amps.

Together, $I_{B1}{}^-$ and $I_{B2}{}^-$ produce the final components of the circuit's overall V_{OSO2} output offset. First, the added $V_{OSO1} = R_1 I_{B1-}$ term also develops a component of V_{OSO2}. As a part of V_{OSO1}, this term defines a current equal to I_{B1-} in R_{1b} for the typical case where $R_{1b} = R_{1a}$. Feedback prevents this current from flowing into R_G and instead diverts it to R_{2b} where it develops an output offset component equal to $V_{OSO2} = -R_2 I_{B1-}$. Finally, feedback conditions similarly constrain the flow of I_{B2-} to that same R_{2b} resistor. As previously mentioned, feedback controls the voltage across R_G as $V_{OS1} - V_{OS2}$, and this defines the current in that resistor independent of I_{B2-}. Similarly, the output of A_1 and the inverting input of A_2 control the voltages at the two ends of R_{1b}, already defining the current there. To accommodate these conditions, feedback draws the I_{B2-} current through the R_{2b} resistor, where it develops an output offset component of $V_{OSO2} = R_2 I_{B2-}$. Combining the three offset components defined here and before for V_{OSO2} produces

$$V_{OSO2} = \left(1 + \frac{R_2}{R_1} + 2\frac{R_2}{R_G}\right)(V_{OS1} - V_{OS2}) + R_2(I_{B2-} - I_{B1-})$$

where $R_{1a} = R_{1b} = R_1$ and $R_{2a} = R_{2b} = R_2$. Compared with the three-op-amp solution at the same gain, this result demonstrates essentially the same offset effect.

7.3.5 CMRR of the two-op-amp configuration

Both the op amps and their feedback resistors potentially affect the common-mode rejection of this instrumentation amplifier. However, in practical cases, just the resistor mismatch determines the final result. Figure 7.10 models the potential common-mode rejection errors with the e_{id} input error signals of the op amps and a consolidated δR_2 resistance error. For the two e_{id} error signals, $CMRR_{OA}$ represents the fundamental common-mode rejection ratio of the associated op amp. The δR_2 error consolidates all of the resistor mismatch effects significant to the circuit's common-mode rejection.

Beginning the analysis with the op amps, the input common-mode signal e_{icm} directly exercises the inputs of A_1 and A_2, producing corresponding input error signals of $e_{id1} = e_{icm}/CMRR_{OA1}$ and $e_{id2} = e_{icm}/CMRR_{OA2}$. In this condition, feedback controls the input circuit in support of these signals to establish the voltages at the inverting inputs of the two op amps. In effect, this feedback state produces an input voltage condition equivalent to that of Fig. 7.6, for which $e_1 = e_{icm} - e_{icm}/CMRR_{OA1}$ and $e_2 = e_{icm} - e_{icm}/CMRR_{OA2}$. Making these substitutions in the e_{o2} expression of that earlier figure produces $e_{o2} - A_{Di}e_{icm}(1/CMRR_{OA2} - 1/CMRR_{OA1})$. This defines an op amp related common-mode gain of

$$A_{CM} = \frac{e_{o2}}{e_{icm}} = A_{Di}\left(\frac{1}{CMRR_{OA2}} - \frac{1}{CMRR_{OA1}}\right)$$

and a common-mode rejection limit of

$$CMRR_{OA} = \frac{A_{Di}}{A_{CM}} = \frac{1}{(1/CMRR_{OA2} - 1/CMRR_{OA1})}$$

For matched op amps, $CMRR_{OA1} \approx CMRR_{OA2}$, making the net $CMRR_{OA}$ infinite. Any residual error resulting from the A_1 and A_2 mismatch typically re-

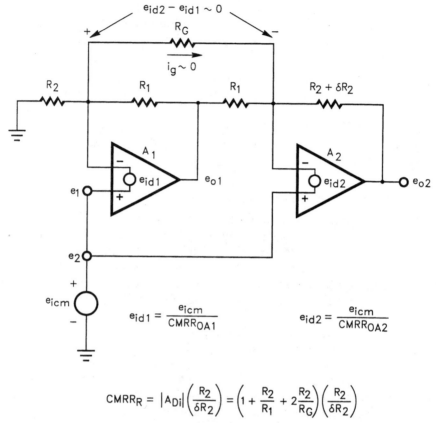

$$\text{CMRR}_R = |A_{Dil}| \left(\frac{R_2}{\delta R_2}\right) = \left(1 + \frac{R_2}{R_1} + 2\frac{R_2}{R_G}\right)\left(\frac{R_2}{\delta R_2}\right)$$

Figure 7.10 Further analysis of the two-op-amp configuration reveals a cancellation of op amp error effects that largely restricts common-mode rejection error to resistor mismatch.

mains insignificant in comparison to the common-mode rejection error produced by δR_2.

The latter error results from the flow of common-mode current through the circuit's feedback resistors. A simplification eases the analysis of this effect since the e_{id} error signals produce counteracting effects in a net signal of $e_{id1} - e_{id2} \approx 0$ across the R_G gain-set resistor. The very small residual signal voltage developed across R_G produces negligible feedback current and permits neglecting R_G in the analysis of the δR_2 effect. Then, A_1 produces the signal $e_{o1} = (1 + R_1/R_2)e_{icm}$ in response to the e_{icm} signal. This e_{o1} signal drives A_2 as an inverting amplifier having a gain of $-(R_2 + \delta R_2)/R_1$. Simultaneously, e_{icm} directly drives A_2 as a noninverting amplifier having a gain of $1 + (R_2 + \delta R_2)/R_1$. Together, the two input signal drives produce a fortuitously simple output error expression of $e_{o2} = -(\delta R_2/R_2)e_{icm}$. In turn, this expression defines the circuit's common-mode gain as $A_{CM} = e_{o2}/e_{icm} = -(\delta R_2/R_2)$. Then, the definition of the common-mode rejection ratio, CMRR = $|A_D/A_{CM}|$, produces the CMRR equation for the two-op-amp instrumentation amplifier. Over the amplifier's useful frequency range, A_D equals the ideal gain A_{Di} and the resistor mismatch limitation to CMRR becomes

$$\text{CMRR}_R = |A_{Di}|\left(\frac{R_2}{\delta R_2}\right) = \left(1 + \frac{R_2}{R_1} + 2\frac{R_2}{R_G}\right)\left(\frac{R_2}{\delta R_2}\right)$$

This expression reveals the primary advantage of the two-op-amp instrumentation amplifier through its high-frequency common-mode rejection. In Sec. 7.2.3, the analogous expression for the three-op-amp configuration displayed common-mode rejection error terms resulting both from an op amp and from resistor mismatch. At higher frequencies, the op amp error dominated, rolling off common-mode rejection. For the two-op-amp configuration, the preceding expressions eliminate the op amp effect due to the circuit's inherent cancellation of the associated CMRR_{OA} effects. Only the δR_2 mismatch remains to limit the common-mode rejection, and this effect displays no direct frequency dependence. Note that δR_2 models the consolidated mismatch error of the circuit's two $R_1{:}R_2$ resistor ratios. Hence, the use of 0.1% tolerance resistors with the circuits would produce a worst-case error of $\delta R_2 = 0.004 R_2$ and a common-mode rejection of CMRR = $250 A_{Di}$.

7.4   Differential-Output Instrumentation Amplifiers

Differential-output signals continue the CMRR benefit of signal transmission as described in Sec. 6.1. There, differential-output transmission on long lines permits the use of common-mode rejection on the receiving end to again remove the coupled noise and ground-potential errors of remote transmission. However, the basic differential-output amplifier described there continues to transmit the common-mode component of the input signal. That component reacts with line impedance imbalances to produce a differential error signal, degrading the common-mode rejection performance as described in Sec. 6.2.1. Section 6.2 presents a common-mode feedback solution to this problem. However, that solution restricts the common-mode input range and signal bandwidth. Instead, adding a second difference amplifier to the three-op-amp instrumentation amplifier produces a differential output that virtually removes these limitations. This alternative offers differential inputs and outputs for common-mode rejection at both ends of the signal transmission without continuing the common-mode signal and without performance-limiting internal feedback. In addition, this alternative offers greater output swing, slew rate, bandwidth, and common-mode rejection than the three-op-amp instrumentation amplifier.

7.4.1 Response of the differential-output configuration

Figure 7.11 shows this configuration with the added difference amplifier formed using A_4 and its associated feedback resistors. The inputs of the circuit's two difference amplifiers make opposite polarity connections to the outputs of A_1 and A_2 to develop the differential output. With its conventional input connections, the difference amplifier

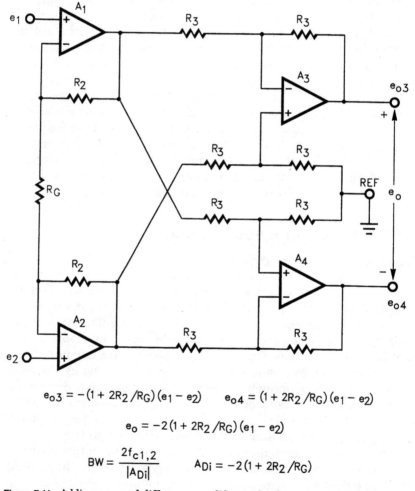

$$e_{o3} = -(1 + 2R_2/R_G)(e_1 - e_2) \qquad e_{o4} = (1 + 2R_2/R_G)(e_1 - e_2)$$

$$e_o = -2(1 + 2R_2/R_G)(e_1 - e_2)$$

$$BW = \frac{2f_{c1,2}}{|A_{Di}|} \qquad A_{Di} = -2(1 + 2R_2/R_G)$$

Figure 7.11 Adding a second difference amplifier to the three-op-amp instrumentation amplifier produces a differential output, as well as increased output swing, slew rate, bandwidth, and CMRR.

formed with A_3 produces the ground-referenced output signal developed in Sec. 7.2.1 of $e_{o3} = -(1 + 2R_2/R_G)(e_1 - e_2)$. Having opposite polarity input connections, the difference amplifier formed with A_4 produces the inverted equivalent signal, or $e_{o4} = (1 + 2R_2/R_G)(e_1 - e_2)$. Together, e_{o3} and e_{o4} develop the final differential output signal $e_o = e_{o3} - e_{o4}$, or

$$e_o = -2\left(1 + 2\,\frac{R_2}{R_G}\right)(e_1 - e_2)$$

This analysis assumes the unity gain commonly used for the difference amplifiers of three-op-amp instrumentation amplifiers, and there the unity-gain choice permits an overall instrumentation amplifier gain also as low as unity. However, the output swing doubling of the differential-output case here makes the minimum-gain magnitude equal to two when $R_G \to \infty$. To restore the unity-gain minimum in this case, the difference amplifier gains would be set at one-half.

However, leaving the difference amplifier gains at unity for the moment permits a more direct comparison of the differential-output and three-op-amp instrumentation amplifiers. This comparison shows that the differential-output alternative doubles three performance ranges. First, output voltage swing and slew rate double due to the two amplifiers that now produce the circuit's output signal. As with the three-op-amp case, the output voltage swing remains limited by difference amplifier output saturation levels. However, unlike the three-op-amp case, the two difference amplifiers here deliver a differential output from opposite-polarity swings that double the normal limit. As a result, the peak-to-peak output voltage range of the differential-output instrumentation amplifier routinely exceeds the net magnitude of the power supply voltages. For comparison, using a single difference amplifier capable of ± 13-V output swing when powered from ± 15-V supplies produces a 26-$V_{p\text{-}p}$ output from a net 30-V supply. However, the two difference amplifiers of the differential-output instrumentation amplifier produce a ± 26-V or 52-$V_{p\text{-}p}$ output swing on the same 30-V supply. This increased swing doubles the output slew rate as well. While one difference amplifier slews in one direction, the other slews in the opposite direction, producing a differential output with twice the slew rate of a single-ended output.

In a third improvement, the differential-output instrumentation amplifier doubles the bandwidth. As described in Sec. 7.2.2, the demand for input precision generally introduces the circuit's bandwidth limit in the differential noninverting amplifier formed with A_1 and A_2. This produces BW $= f_{c1,2}/|A_{Di}|$ for the three-op-amp instrumentation amplifier, where A_{Di} represents the circuit's ideal differential gain of

$A_{Di} = -(1 + 2R_2/R_G)$ and $f_{c1,2}$ represents the unity-gain crossover frequency of A_1 and A_2. As typical of op amp circuits, this bandwidth limit results from the gain demand placed upon A_1 and A_2 and the op amps' constant gain–bandwidth products. For this differential-output case, the preceding e_o expression defines an ideal gain of

$$A_{Di} = -2\left(1 + 2\,\frac{R_2}{R_G}\right)$$

Thus the differential-output instrumentation amplifier delivers twice the gain of the three-op-amp configuration. From another perspective, the differential alternative demands only one-half the gain from A_1 and A_2 to produce the same overall circuit gain. This reduced gain demand doubles the bandwidth delivered by A_1 and A_2 for the differential-output instrumentation amplifier,

$$\text{BW} = \frac{2f_{c1,2}}{|A_{Di}|}$$

where $A_{Di} = -2(1 + 2R_2/R_G)$ and $f_{c1,2}$ represents the unity-gain crossover frequency of A_1 and A_2.

7.4.2   CMRR of the differential-output configuration

In a final improvement, the differential-output instrumentation amplifier improves the high-frequency common-mode rejection dramatically. For the three-op-amp instrumentation amplifier of Sec. 7.2.4 two factors determined common-mode rejection performance. Resistance mismatch dominated the common-mode rejection limit at lower frequencies and the difference amplifier's op amp dominated at higher frequencies. At higher frequencies, the roll off of that op amp's own common-mode rejection response produced an increasing error that surpassed the limit imposed by resistor mismatch. With the differential-output instrumentation amplifier, two difference amplifiers deliver the output signal, and the common-mode rejection errors of their op amps produce counteracting effects for greatly improved high-frequency common-mode rejection.

All four op amps and their feedback resistors potentially affect the common-mode rejection of this instrumentation amplifier. However, in practical cases, just the resistor mismatch of the difference amplifiers dominates the final result over most of the circuit's useful frequency range. First, a qualitative evaluation of common-mode rejection limitations yields performance insight and then analysis quantifies the actual result. Figure 7.12 models the potential common-mode rejection errors with e_{id} input error signals for each op amp and a consolidated

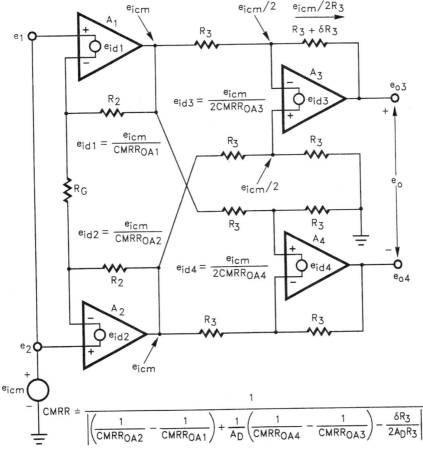

Figure 7.12 Differential-output instrumentation amplifier increases high-frequency CMRR through the counteracting effects of the circuit's two difference amplifiers.

δR_3 resistance error for the difference amplifiers. For each e_{id} error signal, CMRR_{OA} represents the fundamental common-mode rejection ratio of a given op amp. The δR_3 error combines all of the resistor mismatches significant to the circuit's common-mode rejection. Beginning the analysis with the input amplifiers, the full input common-mode signal e_{icm} exercises the inputs of A_1 and A_2, producing corresponding input error signals e_{id1} and e_{id2}. However, for matched input op amps, these error signals produce counteracting effects in a net signal of $e_{id2} - e_{id1} \approx 0$ across the R_G gain-set resistor. Any residual error resulting from a mismatch between CMRR_{OA1} and CMRR_{OA2} typically remains secondary at frequencies below the amplifier's response roll off. Further, the very small residual signal voltage developed across R_G produces negligible feedback current for these amplifiers. As a result, any mismatch in the circuit's R_2 resistors also produces no significant degradation of the common-mode rejection degradation.

While the input amplifiers produce little common-mode rejection error, A_1 and A_2 transfer the common-mode rejection task to the circuit's difference amplifiers. Because e_{icm} produces no feedback currents for A_1 and A_2, these op amps act as voltage followers to that signal and transfer e_{icm} to their outputs. There, e_{icm} drives the inputs of the difference amplifiers and produces the e_{id3} and e_{id4} input error signals. Also, the transferred e_{icm} signal develops a feedback current that reacts with the resistance error δR_3. Combined, the op amp and resistance errors described define the common-mode rejection performance of the differential-output instrumentation amplifier.

Analysis quantifies the result through the definition of the common-mode rejection ratio, CMRR $= |A_D/A_{CM}|$, where A_D and A_{CM} represent the circuit's differential and common-mode gains. From Sec. 7.4.1, the circuit produces a differential gain of $A_{Di} = -2(1 + R_2/R_G)$ over the circuit's useful frequency range. Five error sources define the circuit's common-mode gain, as separated into three components using superposition. First, consider the common-mode rejection effects of the op amp error signals e_{id1} and e_{id2}. Normally, these e_{id} signals include the effects of all input-referred error sources,[5] but this analysis only considers the common-mode rejection error, making $e_{id1} = e_{icm}/\text{CMRR}_{OA1}$ and $e_{id2} = e_{icm}/\text{CMRR}_{OA2}$. These error signals shift the inverting inputs of A_1 and A_2 just as if the ideal circuit of Fig. 7.11 had input signals of $e_1 = -e_{id1}$ and $e_2 = -e_{id2}$. Thus these errors produce an output signal of $e_o = A_D(e_{id2} - e_{id1}) = A_D e_{icm}(1/\text{CMRR}_{OA2} - 1/\text{CMRR}_{OA1})$, reflecting a common-mode gain due to input amplifier errors of $A_{CMi} = e_o/e_{icm} = A_D(1/\text{CMRR}_{OA2} - 1/\text{CMRR}_{OA1})$.

Next, the e_{id} errors of the two difference amplifiers produce an A_{CMd} component of common-mode gain. As described before, the circuit transmits the full e_{icm} signal of the e_1, e_2 input circuit to the outputs of A_1 and A_2. Ideally, the following difference amplifiers remove this signal through subtraction. Common-mode rejection errors limit the effectiveness of this removal through the finite common-mode rejections of A_3 and A_4 and the mismatches of the R_3 resistors. First, consider the op amp errors represented by e_{id3} and e_{id4}. These errors result from the common-mode signal coupled to the inputs of A_3 and A_4. Voltage dividers formed by R_3 resistors reduce this signal to $e_{icm}/2$, giving $e_{id3} = e_{icm}/2\text{CMRR}_{OA3}$ and $e_{id4} = e_{icm}/2\text{CMRR}_{OA4}$. To define the effects of these error signals, consider the A_3 case with the e_{icm} signals indicated. The resulting e_{id3} error signal shifts the inverting input of A_3 just as if a signal of $-e_{id3}$ were impressed at that op amp's noninverting input. Neglecting δR_3 for the moment, this condition produces an output signal of $e_{o3} = -2e_{id3}$. The analogous analysis for A_4 produces $e_{o4} = -2e_{id4}$

for a net differential output of $e_o = e_{o3} - e_{o4} = 2(e_{id4} - e_{id3})$. Substituting for e_{o3} and e_{o4} produces $e_o = e_{icm}(1/\text{CMRR}_{OA4} - 1/\text{CMRR}_{OA3})$, making $A_{CMd} = e_o/e_{icm} = (1/\text{CMRR}_{OA4} - 1/\text{CMRR}_{OA3})$.

Finally, the consolidated δR_3 resistor mismatch error defines the third component of common-mode gain. That resistance error produces a differential output error in response to the feedback current produced by the common-mode signal. As illustrated, e_{icm} transfers directly to the outputs of A_1 and A_2 but only one-half of that signal reaches the noninverting input of A_3. The A_3 feedback replicates the $e_{icm}/2$ signal at this op amp's inverting input, defining the current in the upper left R_3 resistor as $e_{icm}/2R_3$. That current flows through δR_3 to produce an output error signal of $e_o = -e_{icm}\delta R_3/2R_3$ and defines a resistor mismatch component of common-mode gain, $A_{CM} = e_o/e_{icm} = -\delta R_3/2R_3$.

Combining the A_{CMi}, A_{CMd}, and A_{CMr} components yields the circuit's net common-mode gain to define the common-mode rejection ratio. The combination produces

$$A_{CM} = A_D\left(\frac{1}{\text{CMRR}_{OA2}} - \frac{1}{\text{CMRR}_{OA1}}\right)$$

$$+ \left(\frac{1}{\text{CMRR}_{OA4}} - \frac{1}{\text{CMRR}_{OA3}}\right) - \frac{\delta R_3}{2R_3}$$

where CMRR_{OA} represents the common-mode rejection ratio of an individual op amp and $\delta R_3/R_3$ represents the net mismatch of the difference amplifiers' eight resistors. Combining this A_{CM} result with the CMRR definition, $\text{CMRR} = |A_D/A_{CM}|$ defines the common-mode rejection of the differential-output instrumentation amplifier as

$$\text{CMRR} = 1\bigg/\left[\left(\frac{1}{\text{CMRR}_{OA2}} - \frac{1}{\text{CMRR}_{OA1}}\right)\right.$$

$$\left. + \frac{1}{A_D}\left(\frac{1}{\text{CMRR}_{OA4}} - \frac{1}{\text{CMRR}_{OA3}}\right) - \frac{\delta R_3}{2A_D R_3}\right]$$

As with the three-op-amp case, the effects of CMRR_{OA1} and CMRR_{OA2} subtract to virtually remove the common-mode rejection errors of A_1 and A_2. Unlike that previous case, this expression reveals a primary benefit of the differential-output instrumentation amplifier in the counteracting common-mode rejection errors of A_4 and A_3. As expressed, CMRR_{OA4} and CMRR_{OA3} also subtract to largely remove the op amp effect that previously dominated the high-frequency common-mode rejection ratio for the three-op-amp instrumentation amplifier of Sec. 7.2.

Further evaluation of this equation reveals two significant perspectives on the common-mode rejection of this instrumentation amplifier, separated by frequency and gain ranges. At lower frequencies, the CMRR_{OA} response magnitude of the op amps typically remains large, relegating the control of the common-mode rejection ratio to the resistance mismatch of δR_3, and

$$\text{CMRR} \cong \frac{2A_D R_3}{\delta R_3}$$

where δR_3 represents the consolidated mismatch among the circuit's eight R_3 resistors. For example, selecting 0.1% tolerance for these resistors produces a worst-case error of 0.8% for $\delta R_3 = 0.008R_3$, and a low-frequency common-mode rejection of $\text{CMRR} = A_D/0.004$. Resistor trimming commonly improves this performance by at least an order of magnitude.

However, with a trimmed circuit, the reduced effects of op amp CMRR_{OA} errors can still impose a high-frequency limitation. The $(1/\text{CMRR}_{OA2} - 1/\text{CMRR}_{OA1})$ and $(1/\text{CMRR}_{OA4} - 1/\text{CMRR}_{OA3})$ terms of the CMRR expression largely remove the effects of op amp common-mode rejection errors over most of the amplifier's useful frequency range. However at the frequency extreme, these errors potentially return to prominence and can override the smaller common-mode rejection error produced by a trimmed δR_3. Then, the CMRR_{OA} error of either the input or the difference amplifiers can control the final CMRR result. As expressed before, higher A_D gains subdue the effects of the $(1/A_D)(1/\text{CMRR}_{OA4} - 1/\text{CMRR}_{OA3})$ term, generally relegating the high-frequency CMRR degradation to the $(1/\text{CMRR}_{OA2} - 1/\text{CMRR}_{OA1})$ term. In spite of this residual degradation, the $(1/\text{CMRR}_{OA4} - 1/\text{CMRR}_{OA3})$ subtraction of difference amplifier errors makes the differential-output instrumentation amplifier far superior for high-frequency common-mode rejection.

References

1. J. Graeme, *Optimizing Op Amp Performance*, McGraw-Hill, New York, 1997.
2. Graeme, op cit., 1997.
3. Graeme, op cit., 1997.
4. Graeme, op cit., 1997.
5. Graeme, op cit., 1997.

8

Bridge Amplifiers

The low-level outputs of typical transducers place particularly demanding requirements upon their monitoring amplifiers. Connecting the transducer in a bridge configuration greatly eases these requirements by reducing the signal monitored to its deviation from the quiescent state. However, the bridge connection introduces a common-mode signal, requiring differential measurement of the transducer response. Instrumentation amplifiers most commonly serve this purpose by supplying common-mode rejection to remove this extraneous signal. Alternately, an op amp replaces the instrumentation amplifier, using subtraction to remove the common-mode signal. In either case, the bridge connection of a single transducer introduces nonlinearity to the circuit response. For the instrumentation amplifier case, an alternative current bias of the bridge reduces this nonlinearity by a factor of two. For the op amp case, modifying the bridge connection to form a difference amplifier produces a low-gain but linear response.

Alternately, a variety of two-amplifier bridge circuits produce linearized, high-gain responses through bias or feedback control of the bridge. To linearize the response, each of these circuits eliminates the transducer's influence upon its own bias current. With a fixed current, the bridge-connected transducer produces a voltage signal having direct proportionality to the transducer resistance deviation. The first such circuit adds a second transducer and a bias control op amp that establishes a fixed current for the bridge. The next two circuits accommodate a single transducer and modulate either the voltage or the current bias of the bridge to maintain a fixed transducer current. The final two-amplifier circuit accommodates the single transducer and fixed voltage bias of the basic bridge circuit by using two op amps to produce dual feedback control of the bridge. This combination establishes a fixed transducer current and provides high gain to the transducer signal.

Performance comparisons between the various bridge circuits revolve around a standard form equation expressing circuit response. Each of the bridge circuits to be described produces a response of the form

$$e_o = G\!\left(\frac{\delta R/R}{D}\right)\frac{V_B}{x}$$

That standard equation reflects the circuit's voltage gain, responsivity, and sensitivity in the three terms of a multiplied product. Any bridge circuit response reduces to this standard form when manipulated to segregate the effects of voltage gain, transducer deviation, and bridge bias. The first term of this product, G, reflects the circuit's voltage gain as supplied by one or more amplifiers. Next the term $(\delta R/R)/D$ reflects the circuit's responsivity or the relative response effect to a given transducer deviation δR. Within this term, R represents the quiescent resistance of the transducer and D is a denominator term that potentially modifies the responsivity. Ideally, $D = 1$, leaving the responsivity unchanged by transducer deviations, but simple bridge solutions include a δR term in D that produces nonlinearity. Finally, the V_B/x term reflects the circuit's sensitivity through the fraction of the bias signal V_B converted to output signal in response to a given transducer deviation δR. Here, x represents a constant divider, which the circuits that follow make 4, 2, or 1. Obviously, $x = 1$ reflects the highest response sensitivity by enabling the full bias signal in the response. When manipulating a given bridge circuit's response to the standard form, all constant multipliers must be factored out and included in the V_B/x term.

8.1   Basic Bridge Amplifiers

Both instrumentation amplifiers and op amps serve basic bridge monitor circuits. Instrumentation amplifiers remove the bridge's common-mode signal directly through common-mode rejection and op amps remove this signal through signal subtraction. While a more obvious solution, the instrumentation amplifier produces only one-half the bridge sensitivity as compared to the op amp. However, the instrumentation amplifier delivers twice the range of linear response, one-half the offset error, and readily adapts to single-supply operation. These varied performance factors make the bridge amplifier choice a compromise for specific applications.

8.1.1   Basic instrumentation amplifier solutions

As shown in Fig. 8.1, the most common transducer monitor consists of an instrumentation amplifier connected across the outputs of a

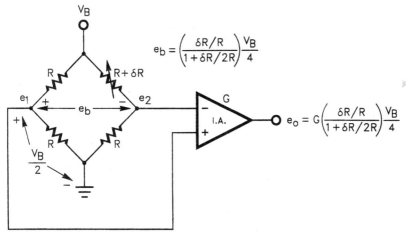

Figure 8.1 Bridge bias of a transducer separates the signal produced by transducer deviation from the quiescent signal but introduces a common-mode signal, requiring differential monitoring.

bridge-connected configuration.[1] In this configuration, the transducer produces a differential output signal e_b corresponding to its δR deviation and independent of its quiescent resistance R. Thus the bridge connection separates the δR signal from its background signal to drive the differential inputs of the instrumentation amplifier. With the otherwise equal resistances of the bridge, this differential signal rides upon a common-mode signal of approximately $V_B/2$, where V_B is the bridge bias voltage. However, the instrumentation amplifier rejects this common-mode signal, extracting the differential signal e_b, and amplifies that differential signal by a gain G to produce the output response $e_o = Ge_b$.

Simple voltage divider analysis quantifies e_b for this expression based upon the ground-referenced e_1 and e_2 output signals of the bridge. From this analysis $e_1 = V_B/2$, $e_2 = (V_B/2)/(1 + \delta R/2R)$, and $e_b = e_1 - e_2$ yields the bridge's differential output signal of

$$e_b = \left(\frac{\delta R/R}{1 + \delta R/2R}\right)\frac{V_B}{4}$$

The instrumentation amplifier amplifies this classic bridge output signal by the gain G to produce the final circuit output $e_o = Ge_b$, or

$$e_o = G\left(\frac{\delta R/R}{1 + \delta R/2R}\right)\frac{V_B}{4}$$

In addition to the e_o expressed, this configuration produces an output offset voltage due to the instrumentation amplifier of $V_{OSO} = GV_{OS}$, where V_{OS} represents the input offset voltage of that amplifier.

As described in the chapter introduction, the equation for e_o reveals the fundamental characteristics of the basic instrumentation amplifier solution in a three-term product expressing gain, responsivity, and sensitivity. Note that the form of the equation moves all constant multipliers or divisors to the $V_B/4$ term, leaving the G term as simply the voltage gain supplied by the instrumentation amplifier and the $(\delta R/R)/(1 + \delta R/2R)$ term as a reduced fraction. The latter term of the product expresses responsivity and reveals the inherent nonlinearity of the single-transducer bridge. That nonlinearity results from the presence of δR in both the numerator and the denominator of the responsivity term. As a result, this fundamental bridge instrumentation only approximates a linear response for small transducer deviations where $\delta R \ll 2R$. In that range,

$$e_o \approx G\left(\frac{\delta R}{R}\right)\frac{V_B}{4}$$

where the e_o responsivity term $\delta R/R$ reflects a direct proportionality to δR.

Both the original and the simplified e_o expressions display proportionalities to $V_B/4$, reflecting the relative sensitivity of the bridge response. For a given transducer deviation δR, the bridge delivers a fraction of the bias voltage V_B as an output signal, and in this case, that fraction displays a proportionality to $V_B/4$ rather than the full V_B. This $V_B/4$ proportionality characterizes the basic bridge having only one active element in its four arms. Other bridge options, not considered here, make all four bridge arms active in a complementary fashion where two transducers produce positive responses to a stimulus while the other two produce negative responses. Such arrangements produce linear responses directly proportional to the total V_B. However, the complementary requirement of such bridges generally restricts their use to lower-accuracy pressure and strain measurements. Alternately, bridge circuits described later employ bias or feedback control of the bridge to improve sensitivity and linearity without the need for four active arms.

8.1.2   Basic op amp solutions

For lower cost and a far greater selection variety, op amps replace the instrumentation amplifier in bridge instrumentation. The basic op amp alternative also doubles measurement sensitivity but degrades linearity and offset performance. With an op amp, signal summation

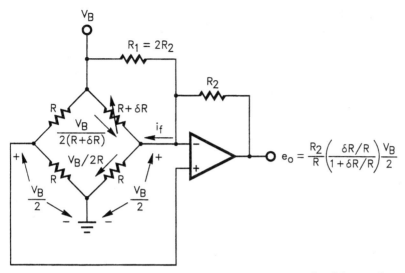

$$e_o = \frac{R_2}{R}\left(\frac{\delta R/R}{1+\delta R/R}\right)\frac{V_B}{2}$$

Figure 8.2 An op amp replaces the common-mode rejection role of the previous instrumentation amplifier with a signal summation that removes the bridge's common-mode voltage from the circuit output.

replaces the instrumentation amplifier's common-mode rejection for removal of the bridges' common-mode output voltage. As shown in Fig. 8.2, this op amp alternative again connects the amplifier's two inputs directly to the bridge outputs. Also as before, the equal resistances of the left side of the bridge produce a voltage of $V_B/2$ at their junction, and that voltage drives the noninverting input of the op amp. To maintain zero voltage between its two inputs, the op amp feedback replicates that voltage at the amplifier's inverting input, making $V_B/2$ a common-mode input signal. That signal would transfer directly to the amplifier output except for the signal summation introduced by R_1. That resistor senses V_B to develop a counteracting output signal component equal to $-(R_2/R_1)V_B$. Making $R_1 = 2R_2$ develops an output component of $-V_B/2$ to cancel the output effect of the op amp's common-mode input voltage. However, this cancellation requires a highly accurate trim of the circuit's R_2/R_1 gain to achieve a common-mode performance comparable to that of the instrumentation amplifier solution. That trim should be performed with the bridge in place and in its quiescent state to compensate also for the common-mode effects of bridge resistance mismatches.

With the common-mode effect canceled, the only signal developed at the circuit output results from the i_f feedback current supplied to the bridge by the op amp. The op amp supplies this current to maintain zero voltage between its two inputs. As described, the bridge bias and the amplifier feedback establish a voltage of $V_B/2$ at these inputs, making $V_B/2$ the voltage across all four arms of the bridge. However, that equal-voltage condition produces unequal currents in the right side of the bridge, as illustrated. The upper right bridge resistor conducts a current of $V_B/2(R + \delta R)$ while the lower right resistor conducts a current of $V_B/2R$. To support this imbalance, feedback from the op amp supplies the current $i_f = V_B/2R - V_B/2(R + \delta R)$ through the R_2 feedback resistor. This develops $e_o = i_f R_2$, or

$$e_o = \frac{R_2}{R}\left(\frac{\delta R/R}{1 + \delta R/R}\right)\frac{V_B}{2}$$

Comparison of this result with the instrumentation amplifier solution of Sec. 8.1.1 reveals the equivalencies, advantages, and disadvantages of the basic op amp alternative. In a first comparison, the result displays a voltage gain of R_2/R, equivalent to the gain G of the previous instrumentation amplifier case. Second, this e_o result expresses a proportionality to $V_B/2$ rather than the previous $V_B/4$ of the instrumentation amplifier solution. That reflects twice the sensitivity to transducer deviation and the principal advantage of the op amp alternative. Next, the e_o expression again contains δR in both its numerator and its denominator, producing a nonlinear response. A linear response only results for small deviations where, for $\delta R \ll R$,

$$e_o \approx \frac{R_2}{R}\left(\frac{\delta R}{R}\right)\frac{V_B}{2}$$

This $\delta R \ll R$ constraint contrasts with the instrumentation amplifier's $\delta R \ll 2R$ requirement, reflecting a 2:1 reduction in the linear range of operation with the op amp solution.

A final comparison shows that the op amp solution also produces at least twice the output offset error. For this comparison, superposition analysis sets $V_B = 0$ and $\delta R = 0$ to produce the offset analysis circuit of Fig. 8.3. There, the op amp appears as a noninverting amplifier to its input offset voltage V_{OS} and amplifies this offset by a gain of $1 + R_2/(R_1\|R\|R)$. Evaluating this offset with the $R_1 = 2R_2$ condition prescribed before yields the output offset expression

$$V_{OSO} = -\left(\frac{3}{2} + 2\,\frac{R_2}{R}\right)V_{OS}$$

As mentioned, the R_2/R term expresses the circuit's voltage gain, equivalent to the gain G of the instrumentation amplifier case. Substituting for comparison produces $V_{OSO} = -(3/2 + 2G)V_{OS}$, and the high gains typically used with bridge amplifiers reduce this expression to $V_{OSO} \approx 2GV_{OS}$, or twice the $V_{OSO} = GV_{OS}$ of the instrumentation amplifier case.

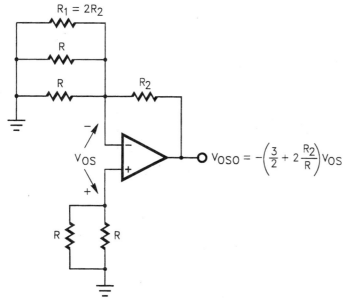

Figure 8.3 Superposition analysis model of Fig. 8.2 shows that the basic op amp alternative for the bridge amplifier increases output offset by at least a factor of two.

8.1.3 Single-supply solutions

Normally, dual-supply amplifiers serve the bridge monitor role even though the bridge itself only requires a single supply. The dual-supply requirement arises from the amplifier's need to operate around zero volts or ground when the transducer deviation approaches zero. However, instrumentation amplifiers permit single-supply operation when combined with an op amp that develops a new common for the circuit.[2] That op amp shifts the circuit's output common reference away from ground to retain linear operation under small or even zero transducer deviation.

Typical instrumentation amplifiers require dual power supplies to support internal bias voltages that would otherwise prevent operation around zero volts. To support these biases, the inputs and outputs of most such amplifiers must operate no closer than about a volt away from the potentials at the two power-supply terminals. Single-supply operation places one of these terminals at zero, requiring offsetting of the amplifier inputs and outputs. Bridge monitoring applications automatically fulfill the input offsetting requirement through the common-mode voltage presented at the bridge outputs. However, the inconvenience of output offsetting remains.

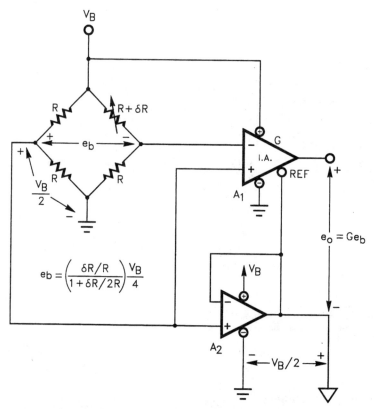

Figure 8.4 For single-supply operation, a voltage follower offsets the instrumentation amplifier reference terminal and provides a new circuit common, permitting differential output measurements.

Instrumentation amplifiers ease the latter task through their common-reference, or REF, terminals. Most applications of these amplifiers connect that terminal to ground but biasing it away from ground directly offsets the output signal. For this purpose, the bridge conveniently generates an offset voltage of $V_B/2$ at the left bridge output, as shown in Fig. 8.4. However, that output cannot be connected directly to the instrumentation amplifier's REF terminal, as the bridge resistance would then degrade the impedance balance required for common-mode rejection. Chapter 7 describes this instrumentation amplifier balance requirement in terms of its internal feedback resistances. To avoid this imbalance, the A_2 voltage follower shown transfers the $V_B/2$ offset to the instrumentation amplifier while presenting a low impedance to the REF terminal. In the process, the $V_B/2$ voltage at the input and at the output of the follower provides offsetting for that amplifier to permit single-supply operation of A_2 as well.

As desired, this configuration offsets the instrumentation amplifier output with respect to the power-supply ground. However, any output signal measurement made with respect to that ground would contain the $V_B/2$ offset voltage, defeating the common-mode rejection function of the instrumentation amplifier. To exclude this offset, differential measurement between the two amplifier outputs excludes this offset in the e_o signal of the figure. Such measurement makes the follower's output a level-shifted common return for all circuit connections following the instrumentation amplifier. That output provides a low-impedance common over a frequency range extending well beyond that exercised by typical bridge signals. However, the current-carrying capacity of this pseudo-common remains limited to the output current capability of the A_2 op amp.

The differential output measurement described reproduces the response result of the basic instrumentation amplifier solution and

$$e_o = G\left(\frac{\delta R/R}{1 + \delta R/2R}\right)\frac{V_B}{4}$$

where G is the voltage gain of the instrumentation amplifier. Refer to Sec. 8.1.1 for discussions of the bridge sensitivity and nonlinearity of this response. The actual differential output signal also contains an offset error due to the input offset voltage V_{OS1} of the instrumentation amplifier. As with the basic case, the instrumentation amplifier amplifies this offset by its gain G to produce $V_{OSO} = GV_{OS1}$. Note that A_2 does not contribute to this offset because the input offset voltage of that amplifier shifts both the REF pin and the pseudo-common, making this offset effect a common-mode signal at the circuit output. The differential output signal e_o remains immune to this common-mode effect.

8.2 Response Linearization with a Single Amplifier

Two modifications to the basic bridge amplifiers of the preceding section improve the response linearity without increasing the number of amplifiers required. For the instrumentation amplifier solution, replacing the voltage bias of the bridge with a current bias reduces the inherent bridge nonlinearity by a factor of two.[3] Next, modifying the bridge connection in the op amp solution converts the circuit to a difference amplifier for complete removal of the inherent nonlinearity. However, this latter solution typically restricts the circuit gain to unity.

8.2.1 Linearized instrumentation amplifier solutions

For the basic instrumentation amplifier solution, current bias of the bridge reduces the response nonlinearity, as illustrated in Fig. 8.5.

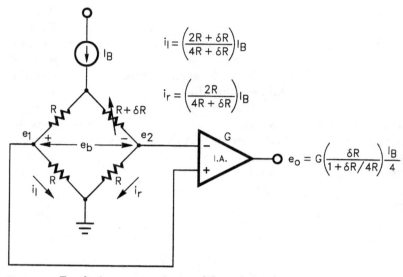

Figure 8.5 For the instrumentation amplifier solution, current bias of the bridge converts the nonlinearity-producing response denominator from $1 + \delta R/2R$ to $1 + \delta R/4R$, doubling the linear range of operation.

Except for the bridge bias source, this circuit duplicates that of Fig. 8.1, and the instrumentation amplifier inputs continue to extract the differential bridge signal e_b from the background of the bridge's common-mode signal. To develop e_b, the left and right sides of the bridge act as a current divider in response to the current bias I_B. The divider action produces left and right side bridge currents of

$$i_l = \left(\frac{2R + \delta R}{4R + \delta R}\right) I_B$$

and

$$i_r = \left(\frac{2R}{4R + \delta R}\right) I_B$$

These currents react with the bridge's lower two R resistances to produce the ground-referenced e_1 and e_2 bridge output signals. The instrumentation amplifier senses these two signals, performs a subtraction to extract the e_b differential signal, and amplifies that to produce the output signal $e_o = G(e_1 - e_2) = Ge_b$, or

$$e_o = G\left(\frac{\delta R}{1 + \delta R/4R}\right)\frac{I_B}{4}$$

Comparison of this result with the earlier response equation of Sec. 8.1.1 reveals the relative merits of current rather than voltage bias of the bridge. First, this equation displays a proportionality to $I_B/4$, reflecting the response sensitivity to transducer deviation. As described before, this proportionality reflects sensitivity in the relative portion of the bias source converted to output signal for a given δR transducer deviation. The $I_B/4$ here compares directly with the earlier $V_B/4$ proportionality, reflecting an equivalent sensitivity. Also as before, the equation expresses a nonlinear response through the presence of δR in both its numerator and its denominator. The circuit produces a linear response only for small transducer deviations, where $\delta R \ll 4R$, that is,

$$e_o \approx G\delta R \frac{I_B}{4}$$

This $\delta R \ll 4R$ restriction reflects the improved linearity provided by the current bias. Previously, the voltage-biased solution of Sec. 8.1.1 produced the linear response restriction $\delta R \ll 2R$, yielding only half the δR range for a given nonlinearity limit. Even greater linearity improvement results from current bias when the bridge employs two transducers, as described in Sec. 8.3.1.

8.2.2   Linearized op amp solutions

For lower-gain applications, a single op amp removes the inherent nonlinearity of a single-transducer bridge totally. This requires a modification of the bridge connection, as shown in Fig. 8.6a, where the transducer returns to the op amp output instead of completing the normal bridge connection. Redrawing this circuit in Fig. 8.6b reveals it to be simply a modified difference amplifier. As described in Sec. 1.3, the basic difference amplifier develops an output signal proportional to the difference between two input signals and rejects any signal common to those inputs. For that amplifier, this common-mode rejection depends critically on the matching of the circuit's two resistor networks.

The configuration in Fig. 8.6b departs from this basic difference amplifier in two ways. The left sides of the input resistors both connect to the same input signal, the bridge bias voltage V_B, suggesting a zero output result. Also, the δR transducer deviation imbalances the resistor network matching required for the basic difference amplifier. However, these deviations produce two fundamental characteristics required of a bridge amplifier, common-mode rejection and a response to transducer deviations. As described in Sec. 8.1.1, a bridge produces a common-mode output signal that must be rejected to permit sensitive detection of the differential signal produced by transducer deviation. Figure 8.6a defines the bridge's common-mode output signal as

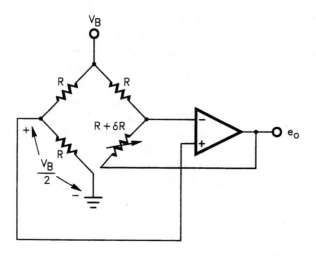

(a)

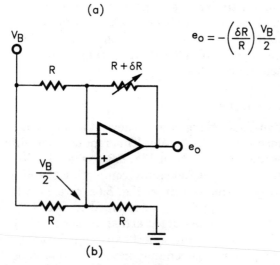

$$e_O = -\left(\frac{\delta R}{R}\right)\frac{V_B}{2}$$

(b)

Figure 8.6 Modifying the bridge connection produces a difference amplifier for a linear, but low-gain response to a transducer using just a single op amp.

$V_B/2$, like that of the basic instrumentation amplifier solution. In its quiescent state with $\delta R = 0$, the bridge still develops this voltage but the circuit rejects it, as best evaluated with Fig. 8.6b. There, the $\delta R = 0$ condition produces balanced resistor networks to reject the effects of the V_B signal applied to both difference amplifier inputs. This produces $e_o = 0$ and illustrates the common-mode rejection feature of the modified difference amplifier.

The condition of $\delta R \neq 0$ would seem to negate this feature, but this very fact serves to develop an output response to the transducer's deviation. By disturbing the circuit's common-mode balance, the modified difference amplifier produces an output signal linearly proportional to δR. Two signal paths define this response through inverting and noninverting connections to the op amp, as separated using superposition analysis. In that analysis, first consider the V_B connection at the upper input resistor R of Fig. 8.6b. This connection produces an inverting amplifier response to V_B and an e_o signal component of

$$e_{oi} = -\left(1 + \frac{\delta R}{R}\right)V_B$$

Next, the V_B connection to the lower input resistor R develops the common-mode voltage $V_B/2$ at the op amp's noninverting input. To this signal, the circuit appears as a noninverting amplifier having a gain of $2 + \delta R/R$ and results in an e_o component of

$$e_{on} = \left(1 + \frac{\delta R}{2R}\right)V_B$$

Combining these two components of e_o defines the net output signal as $e_o = e_{oi} + e_{on}$, or

$$e_o = -\left(\frac{\delta R}{R}\right)\frac{V_B}{2}$$

Comparison of this response with those of previous bridge amplifiers illustrates the advantages and disadvantages of the single op amp solution. The advantages appear in the circuit's linearity and sensitivity. First, the e_o output signal expressed responds to the transducer deviation δR in a direct proportionality without the nonlinearity common to the basic bridge response. All bridge amplifiers previously described produce δR terms in both the numerators and the denominators of their response equations, reflecting nonlinear responses. Next, the e_o response displays a proportionality to $V_B/2$, reflecting a 2:1 improved sensitivity to transducer deviation. Previously, the basic instrumentation amplifier solution of Sec. 8.1.1 produced a corresponding proportionality to $V_B/4$, reflecting only one-half the sensitivity expressed here.

The disadvantages of this op amp solution occur in its lower signal gain and higher offset gain. This e_o expression lacks the gain term G of previous cases which expressed the voltage gain supplied by the amplifier to the bridge output signal. In this modified op amp case, $G = 1$ and this single-amplifier solution only serves applications where low gain remains acceptable. The low-level output signals of most bridges would generally require additional amplification, in which case one of the two-amplifier solutions described in the following sections might better serve the application. Further, the unity-gain difference amplifier formed by the configuration here amplifies the op amp's input offset voltage by a gain of two. This produces an output offset of $V_{OSO} = 2V_{OS}$, where V_{OS} represents the input offset voltage of the op amp. Thus the configuration produces a 2:1 disadvantage in the gain supplied to the op amp offset voltage. Still, the 2:1 increase in bridge sensitivity, described here, maintains the same signal-to-offset ratio in the final output signal.

8.3   Response Linearization through Bias Control

The inherent nonlinearity of the single-transducer bridge results from the transducer's influence upon its own current. With the basic bridge, a fixed bias voltage results in a current through the transducer that equals the bias voltage divided by the resistance of the transducer side of the bridge. For Fig. 8.1, this current equals $V_B/(2R + \delta R)$, which displays a dependence on the transducer deviation δR. The product of this current and the transducer resistance defines the signal voltage of the transducer. Since both terms of this product vary with δR, the result produces a nonlinear response to the transducer deviation. Three bridge amplifier alternatives remove this nonlinearity by controlling the bridge bias to maintain a constant current through the transducer. The first employs a current rather than a voltage bias source for the bridge and adds a second transducer to maintain equal resistances on the two sides of the bridge. Then, the constant bias current divides equally between the two sides to make the transducer current also constant. The second alternative modulates the voltage bias supplied to a bridge to provide a compensating current adjustment in response to a transducer deviation. The third alternative modulates a bridge bias current for the same result.

8.3.1   Linearization with current bias and dual transducers

Biasing a dual-transducer bridge with a current source removes the inherent nonlinearity of the basic bridge. However, reference current sources suitable for this approach remain much more limited in variety than their voltage counterparts. Of course, current sources can be built starting with a voltage reference,[4] but a more direct approach uses op amp control of the bridge to produce this voltage-to-current conversion. Figure 8.7 illustrates this control along with the basic instrumentation amplifier monitor of the bridge. There, the output volt-

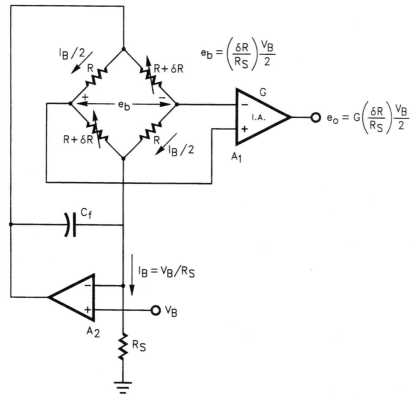

Figure 8.7 Dual transducers and op amp control of current bias remove the inherent nonlinearity of single-transducer voltage-biased bridges.

age of the A_2 op amp replaces the normal voltage bias at the top of the bridge. This output voltage responds to those at the op amp's two inputs, as presented by the reference voltage V_B and the voltage developed by the bridge current I_B across sense resistor R_S. The op amp feedback equalizes its two input voltages by adjusting its output voltage to control I_B. At equilibrium, this control produces a constant bridge bias current of $I_B = V_B/R_S$. Note, however, that this circuit includes the bridge in the op amp's feedback path and connection to remote transducers can introduce parasitic capacitances and inductances that could compromise stability. In such cases, adding the C_f bypass capacitor shown ensures a controlled feedback factor[5] at higher frequencies to restore stability.

With current bias, the dual-transducer bridge linearizes the circuit response by maintaining equal and constant currents in the left and right sides of the bridge. Having two transducers, the two sides act as a fixed current divider in response to the I_B bias. Previously the single-transducer bridge of Sec. 8.2.1 resulted in a variable current divider since transducer variations imbalanced the relative resistances of the bridge's two sides. That produced a varying current division which combined with the varying transducer resistance to produce nonlinearity. In this case, equal resistances on the left and right sides of the bridge ensure that constant currents of $I_B/2$ flow in each side. As a result, the bridge develops a differential output voltage of $e_b = (I_B/2)(R + \delta R) - (I_B/2)R$, where $I_B = V_B/R_S$. Then, $e_b = (\delta R/R_S)\,(V_B/2)$ and the gain supplied by the instrumentation amplifier produces the output result in Fig. 8.7 of

$$e_o = G\left(\frac{\delta R}{R_S}\right)\frac{V_B}{2}$$

Examination of this result reveals two advantages of the current-biased dual-transducer bridge. First, output signal e_o responds to the transducer deviation δR only through a numerator term, reflecting a linear response. Previously, other bridge instrumentation solutions resulted in δR in both the numerators and the denominators of their response expressions, reflecting nonlinearities. The dual transducers also double the bridge's sensitivity, as reflected by the $V_B/2$ term of this response. This measure of sensitivity reflects the portion of the bias signal converted to an output signal in response to a given transducer deviation. Previously, the basic instrumentation amplifier solution of Sec. 8.1.1 produced a response sensitivity proportional to $V_B/4$, or one-half that achieved here.

Amplifier input offset voltages introduce errors in this output response in terms of both an output offset voltage and a sensitivity deviation. As before, the A_1 instrumentation amplifier amplifies its input offset voltage along with the bridge signal to produce an output offset voltage of $V_{OSO} = GV_{OS1}$. The A_2 op amp does not contribute to V_{OSO} but instead alters the bridge sensitivity by modifying the $V_B/2$ term of the e_o expression. As described before, the op amp feedback ideally transfers the bias voltage V_B to the R_S sense resistor, establishing the constant bridge bias current. The input offset voltage of A_2 detracts from this ideal transfer, replacing the $V_B/2$ term of the e_o expression with $(V_B - V_{OS2})/2$.

8.3.2 Linearization through voltage bias modulation

Signal-dependent modulation of a bridge bias voltage also removes the inherent nonlinearity without the need for the second transducer of the preceding solution. The response equation of the basic instrumentation amplifier solution of Sec. 8.1.1 expresses this linearization opportunity. There,

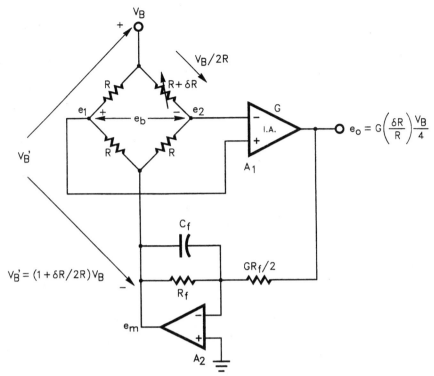

Figure 8.8 Modulation of the bridge bias voltage through an inverting feedback amplifier removes the inherent nonlinearity of a single-transducer bridge.

$$e_o = G\left(\frac{\delta R/R}{1 + \delta R/2R}\right)\frac{V_B}{4}$$

The nonlinearity of this response develops due to the presence of δR in its denominator, but making V_B proportional to $(1 + \delta R/2R)$ removes that effect. This compensating adjustment requires increasing V_B in response to an increase in δR and, fortuitously, the output e_o expressed here presents a signal with fundamentally the correct proportionality.

Figure 8.8 utilizes and modifies this signal to produce exactly the required bias modulation through the addition of an inverting amplifier. That amplifier, formed with A_2, senses the output signal of the A_1 instrumentation amplifier, amplifies that signal by a gain of $-2/G$, and drives the normally grounded point of the bridge with a modulation signal. That signal, $e_m = -2e_o/G$, increases the instantaneous bias voltage across the bridge to $V'_B = V_B + 2e_o/G$ and produces the ground-reference bridge output signals $e_1 = V'_B/2$ and $e_2 = RV'_B/(2R + \delta R)$. This combination develops the bridge's differential output signal of $e_b = e_1 - e_2$, which the instrumentation amplifier amplifies to produce the final output signal $e_o = G[(\delta R/R)/(1 + \delta R/2R)](V'_B/4)$. Expressed this way, the new e_o result displays the same form as the basic instrumentation amplifier result repeated earlier. However, in this case the variable V'_B replaces the previous constant V_B, and the substitution of $V'_B = V_B + 2e_o/G$ yields the linearized output expression for Fig. 8.8,

$$e_o = G\left(\frac{\delta R}{R}\right)\frac{V_B}{4}$$

Further analysis lends insight into the linearity correction achieved beginning with a final solution for V'_B. From before, $V'_B = V_B + 2e_o/G$ and substituting the preceding e_o expression reduces this to $V'_B = (1 + \delta R/2R)V_B$. This V'_B result displays the exact proportionality required for linearity correction as described at the beginning of this section. From another perspective, the inherent nonlinearity of a single-transducer bridge results from the transducer's influence upon its own current. With a fixed bias voltage, a transducer resistance change in that simple bridge simultaneously modifies the transducer current such that the product of the resistance and the current produces a nonlinear response. With the modulated bridge bias here, the current through the transducer remains fixed at $V'_B/(2R + \delta R) = V_B/2R$, avoiding that nonlinearity.

Three other characteristics of this modulated-bias circuit define its residual nonlinearity, sensitivity, and frequency stability. As described, the circuit produces exact cancellation of the bridge's inherent nonlinearity when the A_2 circuit provides a gain equal to $-2/G$, where G is the gain of the A_1 instrumentation amplifier. In practice, the circuit's two amplifiers do not achieve an exact gain ratio match and the residual error limits the accuracy of the linearization. However, a practical 0.1% ratio match reduces the bridge's inherent nonlinearity by 1000:1, leaving the residual nonlinearity a function of the transducer's actual response. Next, the output expression $e_o = G(\delta R/R)(V_B/4)$ expresses the circuit's relative sensitivity to transducer deviation in the $V_B/4$ term, and this matches the corresponding term for the basic instrumentation amplifier solution of Sec. 8.1.1. However, other circuits, also described earlier, produce a proportionality to $V_B/2$, reflecting twice the sensitivity. Finally, this circuit's frequency stability depends on the response characteristics of two amplifiers connected in a common feedback loop. For stability, one of the two must produce a dominant pole, as set by the feedback capacitor C_f added to the A_2 circuit. Stability analysis[6] guides the selection of this capacitor for a given set of application conditions.

Amplifier input offset voltages add error terms to the actual e_o response of the circuit. As always, these offset voltages produce an output offset, but in addition, this circuit's feedback produces an associated sensitivity error as well. Including the V_{OS1} and V_{OS2} input offset voltages of the A_1 and A_2 amplifiers in the previous response analysis produces

$$e_o = G\left(\frac{\delta R}{R}\right)\frac{V_B + 2V_{OS1} + (1 + 2/G)V_{OS2}}{4} + GV_{OS1}$$

where G represents the gain of the instrumentation amplifier.

Comparing this offset-influenced response with the previous e_o result reveals the effects of the input offset voltages of the circuit's two amplifiers. This comparison first shows that the amplifier offsets do not alter the response's simple proportionality to $\delta R/R$, so the offsets do not degrade linearity. However, as would be expected, the input offset voltage of the A_1 instrumentation amplifier produces an output offset error reflected by the added term GV_{OS1}. Note that the V_{OS2} input offset of A_2 does not contribute to this output offset. However, both amplifier offsets produce response sensitivity errors reflected by the offset terms added to V_B. In the absence of these offsets, the previous result expressed a circuit sensitivity proportional to $V_B/4$ alone. For this sensitivity figure of merit, including the amplifier offset effects replaces V_B with $V_B + 2V_{OS1} + (1 + 2/G)V_{OS2}$, as expressed before. However, these offset effects typically remain small in comparison with the tolerance error of the V_B bias voltage.

8.3.3 Linearization through current bias modulation

Signal modulation of current rather than voltage bias of the bridge also removes the inherent nonlinearity of the single-transducer bridge. In Sec. 8.2.1, basic current bias reduced but did not remove this nonlinearity for that bridge configuration. In Sec. 8.3.1, the addition of a second transducer totally removed the inherent nonlinearity. However, the cost of this second transducer often outweighs that of an improved instrumentation solution. Feedback modulation of the bridge bias current provides this improvement, as illustrated in Fig. 8.9. There, the typical instrumentation amplifier monitors the bridge's differential output voltage e_b to produce a buffered output signal directly proportional to the transducer deviation δR.

In this circuit, current bias for the bridge results from the feedback control of the A_2 op amp and a feedback current supplied by the A_1 instrumentation amplifier. Having a grounded noninverting input, the A_2 op amp provides the primary control of this current by forcing the voltage at the lower end of the R_S sense resistor to zero. To do so, the

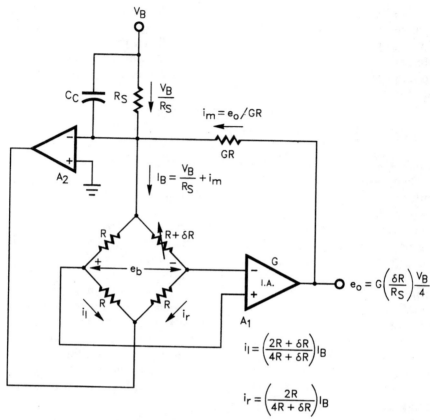

Figure 8.9 Op amp feedback control of basic bias and a modulation feedback current i_m from the instrumentation amplifier combine to bias the bridge with a linearity-correcting current I_B.

op amp output drives the bottom point of the bridge to whatever voltage required to make the current in R_S equal V_B/R_S. That current biases the bridge in conjunction with the feedback modulation current i_m supplied by the A_1 instrumentation amplifier. The latter amplifier supplies i_m through the scaled feedback resistance GR, where G represents the voltage gain of the instrumentation amplifier. The virtual ground produced by A_2 sets the left side of the GR resistance at zero volts to make this modulation current $i_m = e_o/GR$. Together, the two amplifiers produce a net bridge bias current of $I_B = V_B/R_S + i_m = V_B/R_S + e_o/GR$. To this I_B current, the bridge acts as a current divider and produces left and right side bridge currents of

$$i_l = \left(\frac{2R + \delta R}{4R + \delta R}\right) I_B$$

and

$$i_r = \left(\frac{2R}{4R + \delta R} \right) I_B$$

As illustrated, these two currents flow through the bridge's lower resistances R to produce the differential output voltage $e_b = (i_l - i_r)R$. The instrumentation amplifier amplifies the resulting e_b by the gain G to produce the net output signal for the circuit, and combining the various equations presented for i_l, i_r, e_b, and I_B defines this signal as

$$e_o = G\left(\frac{\delta R}{R_S} \right) \frac{V_B}{4}$$

Thus the modulated current bias of a single-transducer bridge results in an output signal linearly proportional to the transducer deviation δR.

The source of this linearity correction lies in the constant bias current maintained in the transducer. Combining the preceding equations for i_r, I_B, and e_o shows this current to be $i_r = V_B/2R_S$. The accuracy of the linearity correction achieved with this approach depends on matching the GR feedback resistance with the product of the instrumentation amplifier gain G and the bridge quiescent resistance R. Also as expressed in e_o, this solution produces a response sensitivity proportional to $V_B/4$, like that of the basic instrumentation amplifier solution, but less than the $V_B/2$ sensitivity indicator of other solutions previously presented.

Amplifier input offset voltages add error terms to the actual e_o response of the circuit. As always, these offset voltages produce an output offset but, in addition, this circuit's feedback produces an associated sensitivity error as well. Including the V_{OS1} and V_{OS2} input offset voltages of the A_1 and A_2 amplifiers in the previous response analysis produces

$$e_o = G\left(\frac{\delta R}{R} \right) \frac{V_B + (R_S/GR)V_{OS1} + (1 + R_S/GR)V_{OS2}}{4} + GV_{OS1}$$

where G represents the gain of the instrumentation amplifier.

Comparing this offset-influenced response with the previous e_o result reveals the effects of the input offset voltages of the circuit's two amplifiers. This comparison first shows that the amplifier offsets do not alter the response's simple proportionality to $\delta R/R_S$, so they do not degrade linearity. However, once again, the input offset voltage of the A_1 instrumentation amplifier produces an output offset error reflected by the added term GV_{OS1}. Note that the V_{OS2} input offset of A_2 does not contribute to this output offset. Still, both amplifier offsets produce response sensitivity errors reflected by the offset terms added to V_B. In the absence of these offsets, the previous result expressed a circuit sensitivity proportional to $V_B/4$ alone. For this sensitivity figure of merit, including the amplifier offset effects mentioned replaces V_B with $V_B + (R_S/GR)V_{OS1} + (1 + R_S/GR)V_{OS2}$. Typically, these offset effects remain small in comparison with the tolerance error of the V_B bias voltage.

As before, this linearity correcting circuit interconnects the feedback paths of two amplifiers, potentially compromising stability. The output signal e_o of the instrumentation amplifier drives the inverting input of the op amp through the GR resistance. In turn, the op amp output drives the noninverting input of the instrumentation amplifier through the bridge to complete a two-amplifier feedback loop. Stabilization of this common loop[7] may require the addition of the C_C phase compensation capacitor shown. That capacitor rolls off the influence of e_o upon the op amp and can introduce a low-frequency dominant pole without restricting the very low-frequency signals typically produced by the bridge.

8.4   Response Linearization through Feedback Control

The preceding bridge linearization circuits combine an instrumentation amplifier and an op amp for bridge bias control. For lower cost, a second op amp replaces the instrumentation amplifier in a configuration that places all four legs of the bridge under feedback control. As shown in Fig. 8.10,[8] this configuration combines the basic transconductance and transimpedance amplifiers of Chap. 4. Op amp A_1 and the bridge form the transconductance amplifier and rotating the bridge as shown eases visual comparison with the circuit of Sec. 4.3. In this rotated position, the normal bridge outputs still connect to the differential inputs of an amplifier. However, in this case, the A_1 op amp replaces the previous instrumentation amplifier. Also, the normally grounded terminal of the bridge now returns to the output of this op amp. As will be seen, this modified return establishes part of the feedback control that linearizes the bridge response. This A_1 transconductance amplifier supplies an output current i_o to the transimpedance amplifier formed with A_2 and R_G, and the latter amplifier completes the feedback control of the bridge.

A largely intuitive evaluation of this circuit's feedback control quantifies the net circuit response. First, the A_2 feedback produces a virtual ground at that op amp's inverting input. This establishes a zero-volt bias at the lower bridge output terminal, and the feedback of A_1

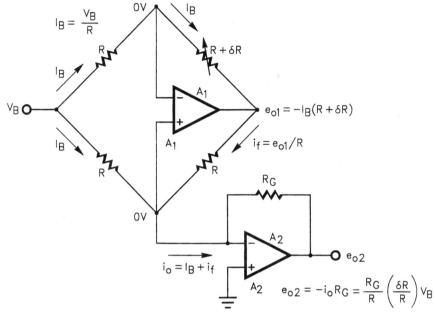

Figure 8.10 Two op amps linearize bridge response with lower cost and greater sensitivity through dual-feedback control of the bridge.

replicates this zero-volt condition at its inverting input, controlling the upper bridge output terminal. With these bias constraints, the entire bridge bias voltage V_B drops across the two left legs of the bridge. The resulting bridge bias currents, $I_B = V_B/R$, define an output current i_o in conjunction with a feedback current supplied by A_1. Flow of the upper I_B current through the transducer produces a voltage drop of $e_{o1} = -I_B(R + \delta R)$ and the A_1 feedback replicates this voltage on the lower right leg of the bridge. That leg conducts a feedback current of $i_f = e_{o1}/R$ which combines with the lower I_B current to produce $i_o = I_B + i_f = -I_B(\delta R/R)$. Finally, the A_2 transconductance amplifier converts this signal current to the output voltage $e_{o2} = -i_o R_G$, or

$$e_{o2} = \frac{R_G}{R}\left(\frac{\delta R}{R}\right)V_B$$

Evaluation of the three primary terms of this response equation reveals the relative merits of this two-op-amp solution. First, R_G/R represents the circuit's voltage gain, replacing the gain G of the previous instrumentation amplifier solutions. Typically, the bridge resistance R remains relatively low, permitting high levels of voltage gain for this two-op-amp circuit. Next, the $\delta R/R$ term of e_{o2} reflects the desired linear response to the transducer's deviation, matching the bias control solutions of the preceding section. Finally, the V_B term expresses an improved sensitivity to that deviation. That sensitivity reflects the fraction of the bridge bias voltage converted to output signal in response to a given transducer deviation δR through the V_B term of the e_o expression. Previous circuits produced response proportionalities to $V_B/4$ or $V_B/2$, reflecting significantly lower sensitivities.

As before, amplifier input offset voltages add error terms to the actual e_o response of the circuit, both in its output offset and in its response sensitivity. Including the V_{OS1} and V_{OS2} input offset voltages of the A_1 and A_2 amplifiers in the previous response analysis produces the circuit voltages and currents illustrated in Fig. 8.11. Combining these signals through the equations shown there produces the offset-adjusted output signal

$$e_o = \frac{R_G}{R}\left(\frac{\delta R}{R}\right)(V_B + V_{OS1} + V_{OS2}) + 2\frac{R_G}{R}V_{OS1}$$

Comparing this adjusted response with the preceding e_o result reveals the effects of the input offset voltages of circuit's two amplifiers. This comparison first shows that the amplifier offsets do not alter the response's simple proportionality to $\delta R/R$, so they do not degrade linearity. However, as would be expected, the input offset voltage of the A_1 amplifier produces an output offset error, reflected by the added term $2(R_G/R)V_{OS1}$. Thus compared with the R_G/R signal gain of the equation, the circuit delivers twice the gain to V_{OS2} as it does to the transducer deviation $\delta R/R$. However, the mentioned increased sensitivity of the circuit maintains a signal-to-offset ratio that matches the best of the alternatives described before. Note that the V_{OS2} input offset of A_2 does not contribute to this output offset. However, both amplifier offsets produce response sensitivity errors reflected by the offset terms added to V_B. In the absence of these offsets, the previous result expressed a circuit sensitivity directly proportional to V_B alone. For this case, including the amplifier offset effects replaces V_B with $V_B + V_{OS1} + V_{OS2}$. However, these offset effects typically remain small in comparison with the tolerance error of the V_B bias voltage.

References

1. J. Graeme, "Boost Bridge Performance with Economical Op Amps," *EDN*, p. 111, August 20, 1978.
2. J. Graeme, "One Supply Powers Precision Bridge Circuit," *EDN*, p. 207, October 3, 1985.
3. J. Graeme, "Tame Transducer Bridge Errors with Op Amp Feedback Control," *EDN*, p. 173, May 26, 1982.

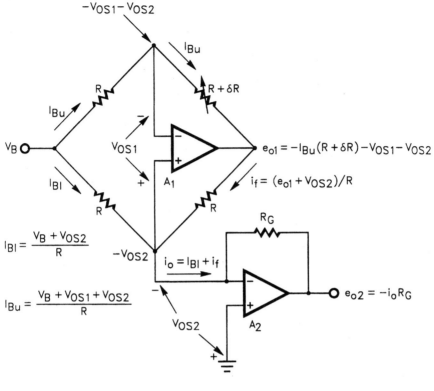

$$e_{o2} = \frac{R_G}{R}\left(\frac{\delta R}{R}\right)(V_B + V_{OS1} + V_{OS2}) + 2\frac{R_G}{R}V_{OS1}$$

Figure 8.11 For the circuit in Fig. 8.10, only A_1 produces an output offset voltage but both amplifiers alter the circuit's sensitivity by adding to the V_B term of the circuit's e_{o2} response.

4. J. Graeme, "Op Amps Turn Voltage References into Current Sources," *EDN*, p. 191, April 26, 1990.
5. J. Graeme, *Optimizing Op Amp Performance*, McGraw-Hill, New York, 1997.
6. Graeme, op cit., 1997.
7. Graeme, op cit., 1997.
8. U.S. patent 4,229,692.

9

Clamping Amplifiers

Clamping amplifiers or feedback limiters provide amplitude control for signal clipping, signal squaring, and overload protection. The clipping and squaring functions provide obvious signal processing functions, and the overload protection prevents measurement delays or the phase reversals that could latch servo loops. To define the clamping voltage level, a zener diode or its bandgap equivalent generally offers the simplest solution for a fixed higher-voltage clamping action of moderate precision. Adding op amps and diode bridges to the clamping circuit improves precision by reducing output impedance, sharpening the clamping transition, balancing clamping symmetry, and increasing speed. Replacing the zener diodes with power-supply-referenced voltage dividers and rectifying junctions expands clamping options to a greater variety and range of voltages. Adding the feedback control of the rectifying junctions sharpens and flattens the clamping response while extending the clamping range and accuracy to very low levels. Finally, replacing the power-supply reference with a control voltage permits electronic variation of clamping levels for test and amplitude modulation requirements.

9.1 Zener-Controlled Clamping

Zener diodes provide the simplest clamping action but with numerous performance limitations. The clamping amplifiers described in this section remove most of these through various additions of a single op amp and a few external components. Adding op amp feedback to the basic zener clamp greatly reduces output impedance to avoid loading errors. Adding decoupling diodes to this feedback sharpens the clamping response by diverting zener leakage current away from the circuit input. Enclosing a single zener diode within a feedback diode bridge improves bipolar clamping symmetry by eliminating the effect of zener

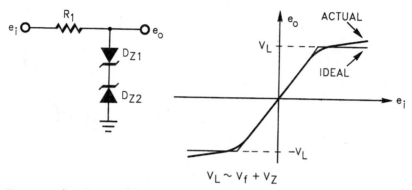

Figure 9.1 A resistor and back-to-back zener diodes provide bipolar limiting but present a high output impedance and produce rounded limit corners followed by sloping limit levels.

mismatch. Finally, providing continuous bias to this clamping zener diode both sharpens the response and increases the clamping speed.

9.1.1 Basic zener clamping

In the simplest clamping structure, zener diodes provide bipolar clamping with two such diodes in a series-opposed connection, as illustrated in Fig. 9.1.[1] For a given signal polarity, the back-to-back D_{Z1} and D_{Z2} zener diodes limit the e_o output voltage with one diode forward-biased and the other in zener breakdown. This produces an output voltage limit of $V_L \approx V_f + V_Z$, where the two components of V_L represent the forward and reverse diode voltage levels. The forward and reverse roles of D_{Z1} and D_{Z2} switch with the reversal of the signal polarity to provide limiting of both positive and negative signals, as illustrated by the response curve in Fig. 9.1. Thus the circuit produces bipolar limiting that occurs symmetrically with respect to ground, to the degree supported by diode matching. Manufacturing process variances inherently mismatch both the V_f and the V_Z diode voltages of D_{Z1} and D_{Z2} to disturb the actual clamping symmetry. In addition, variations in V_f and V_Z with temperature alter the V_L clamping voltage. However, the appropriate choice of the zener diode makes the two thermal variations somewhat counteractive.

While simple in structure, this fundamental limiter suffers from basic performance limitations in the form of high output impedance and low precision. First, in the unlimited state, this simple circuit presents a rather high output impedance, equal to R_1, making the circuit susceptible to loading errors. Second, the zener diode response characteristic compromises clamping precision, as shown by the actual response curve of Fig. 9.1. That curve deviates from the ideal with rounded corners and sloping clamp levels. The rounding results from the zener diode current–voltage characteristic, which lacks the ideal abrupt corner. Continued slopes in the clamp levels result from the zener diode resistance, which forms a voltage divider with R_1. In the clamped state, this divider produces a transfer response attenuation of approximately $e_o/e_i \approx (R_{Zf} + R_{Zr})/R_1$, where R_{Zf} and R_{Zr} represent the forward and reverse resistances of the zener diodes and $R_1 \gg R_{Zf} + R_{Zr}$.

In addition, this fundamental limiter circuit suffers from restricted clamping voltage alternatives, clamping symmetry error, and low speed. First, the V_L voltage realizable with this basic circuit remains limited by the voltage selection variety of zener diodes. This restricts clamping to specific fixed levels and precludes levels below about 1.25 V. Next, this basic circuit produces a clamping symmetry error in its bipolar limiting due to the inherent mismatch between the two zener diodes. Finally, the high capacitance levels common to zener diodes delay the clamping response and reduce the signal bandwidth. Clamping amplifiers described later in this section remove or mitigate most of these limitations using a single op amp.

It should be noted that compensated zener diodes and bandgap equivalents to zeners will not function in circuits that employ two series-opposed connected zener diodes. That diode combination, as shown in Fig. 9.1, requires that a given zener diode act as a shunt regulator under one polarity of applied voltage and as a forward-biased diode under the opposite polarity. Compensated zener diodes include a signal diode in series with the normal zener junction to counteract the thermal drift of the zener voltage. However, that added diode prevents the zener diode from operating as a signal diode under forward-bias conditions. Similarly, bandgap equivalents to zener diodes function as shunt regulators to produce a reference voltage equivalent to that of a zener diode in the reverse or breakdown state. However, these reference devices again do not duplicate the forward-biased performance of the basic zener diode.

9.1.2 Reducing clamping output impedance

The addition of an op amp to the basic bipolar limiter overcomes the output impedance problem, as illustrated in Fig. 9.2.[2] There, an op amp isolates the circuit output from the R_1 resistance by presenting the amplifier's low output impedance at the e_o terminal. This circuit operates as a simple inverting amplifier in both the linear and the clamped states. In the linear region before clamping, the circuit's R_1, R_2 feedback network produces the familiar voltage gain of $-R_2/R_1$. In the clamped states, the circuit limits the output voltage through diode shunting of the R_2 feedback resistor and the virtual ground established at the op amp's inverting input. This combination limits the magnitude of the voltage across R_2, and thereby e_o, to $V_L \approx V_f + V_Z$,

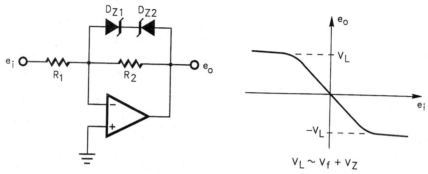

Figure 9.2 Combining an op amp with the basic limiter of Fig. 9.1 produces a clamping amplifier with low output impedance and the option for voltage gain in the unlimited state.

where the two components of V_L represent the forward and reverse diode voltage levels.

While this basic op amp limiter reduces the output impedance, it does not otherwise improve the clamping response and may slightly degrade it through the op amp's offset voltage. In the clamped states, the circuit continues to act as an inverting amplifier, again producing nonzero slopes in the e_o-versus-e_i response shown. This continuation results from the resistances of the zener diodes. Assuming those resistances remain small compared to R_2, they convert the circuit's clamped voltage gain to $e_o/e_i \approx -(R_{Zf} + R_{Zr})/R_1$, where R_{Zf} and R_{Zr} represent the forward and reverse resistances of the zener diodes. Comparing this clamped-state gain with that of the basic zener limiter before shows that the basic op amp limiter offers no improvement in this regard. Also, the rounded voltage–current characteristic of the zener diodes continues to produce soft corners in the clamping response. Added to these response errors, the op amp's input offset voltage shifts the circuit's two clamping levels by $-V_{OS}$. However, for this basic clamping amplifier that offset shift typically remains negligible in comparison with the V_L errors introduced by the zener diodes' voltage tolerances.

9.1.3 Sharpening clamping corners

The rounded corners of the preceding response curves result from the "soft" turn-on characteristics of most zener diodes. In essence, the current–voltage characteristics of these diodes display high levels of leakage current prior to entering the zener breakdown mode. For the preceding circuits, that leakage current begins to limit the output voltage prematurely, producing the rounded response corners of the e_o-versus-e_i plots shown there. To prevent this, a modification to the clamping amplifier of Fig. 9.2 decouples that leakage current from the

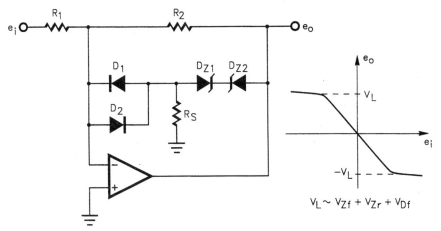

Figure 9.3 Adding decoupling diodes and a shunt resistor to the clamping amplifier of Fig. 9.2 sharpens the clamping corners by diverting the zener leakage current generated prior to breakdown.

amplifier input, as shown in Fig. 9.3.[3] There, the addition of the signal diodes D_1 and D_2 and the shunt resistor R_S protects the amplifier input circuit from the zener leakage effects.

Prior to clamping, the voltage developed across R_2 in this circuit still induces leakage currents in D_{Z1} and D_{Z2}, but these currents no longer couple directly to the amplifier input. Resistor R_S shunts them to ground, and as long as the voltage developed on that resistor remains small, signal diodes D_1 and D_2 remain in the OFF state to block zener current flow to the amplifier input. In this condition, only the small leakage currents generated in D_1 and D_2 reach that input. As the output voltage increases, so does the voltage developed on R_S, and this eventually forward-biases one of the D_1 or D_2 signal diodes to initiate a clamp state. Then, current flow through the feedback diodes diverts that supplied by R_1 to again limit the voltage developed on R_2. As illustrated, this modified action greatly reduces the rounding of the response transition between linear and clamped states. A residual rounding remains, but this reflects the much sharper current–voltage characteristic of the signal diodes D_1 and D_2.

However, the sloping clamp levels produced by the zener diode resistances remain and increase slightly due to the added resistances of the signal diodes. These added diodes also alter the clamp voltage magnitude, making it

$$V_L \approx V_{Zf} + V_{Zr} + V_{Df}$$

where the three components of V_L represent the forward and reverse voltages of the zener diodes and the forward voltage of the signal diodes.

As a side benefit, the decoupling diodes also improve the signal bandwidth. Previously, the circuit of Fig. 9.2 connected the zener diodes directly across the R_2 feedback resistor. Those diodes typically display high levels of junction capacitance that bypass R_2 at higher frequencies. In this case, the inclusion of D_1 and D_2 places the much lower capacitances of the signal diodes in series with those of the zener diodes, and the combined series capacitance produces far less high-frequency shunting of R_2. Section 9.1.5 describes an alternate circuit for sharpening the clamping corners, which similarly improves the bandwidth as well as the clamping symmetry, through the addition of a few more components.

9.1.4 Balancing clamping symmetry

The clamping amplifiers described depend on the matching of two zener diodes to produce symmetry in bipolar limiting. With exact matching, these diodes produce positive and negative clamping levels, both having magnitudes of $V_f + V_Z$, where V_f represents the diodes' forward voltage and V_Z represents the reverse or zener voltage. While their forward voltages tend to match reasonably well, the reverse voltages display a much wider variation and produce asymmetry in the actual output clamping. Replacing the basic two-zener clamp with a single zener and a diode bridge greatly improves symmetry.[4]

As shown in Fig. 9.4, this alternative encloses the zener diode within the bridge in a manner that makes that one zener produce both the positive and the negative output limits. Positive output signals become

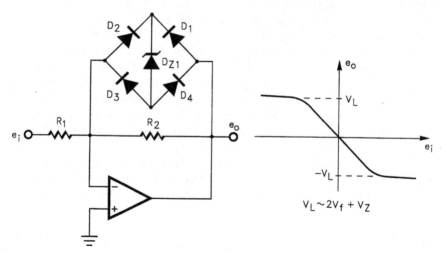

Figure 9.4 Replacing the two zener diodes of preceding clamping amplifiers with one zener diode enclosed in a diode bridge greatly improves the symmetry of bipolar clamping by eliminating zener mismatch error.

clamped when D_1 and D_3 conduct to connect the anode of D_{Z1} to the op amp's inverting input and the cathode to the circuit output. Negative output signals limit when D_4 and D_2 conduct to reverse the connection of the zener diode. Thus clamping occurs when the signal voltage developed across R_2 supports the reverse voltage of the zener diode and the forward voltages of two bridge diodes. For each polarity, this clamping action produces limit levels having magnitudes of

$$V_L \approx 2V_f + V_Z$$

where V_f represents the forward voltage of the bridge diodes and V_Z represents the reverse or zener voltage of D_{Z1}. While this solution introduces a second V_f voltage to V_L, it retains the same V_Z voltage for the two output polarities to improve clamping symmetry significantly. In addition, this solution permits the use of the compensated zener diodes and bandgap equivalents not previously suitable for the series-opposed connection of two such elements.

9.1.5 Improving clamping speed, sharpness, and symmetry

The multiple objectives of this section's title might suggest a profound circuit modification. However, simply providing continuous bias to the zener diode with two resistors produces all three results. As shown in Fig. 9.5, this clamping amplifier alternative differs from the one in Fig. 9.4 only by the inclusion of bias resistors R_3 and R_4. As in the previous case, the bridge connection of a single zener diode again greatly improves clamping symmetry. However, in that case, only the signal current served to bias the zener diode, and a reversal of the signal polarity required the turn-off and then turn-on of the zener diode to alternate clamping polarities. That action requires driving the zener diode through its full voltage–current response. When appropriately chosen, the R_3 and R_4 resistors added here ensure continuous bias of the zener diode to override this on–off action. Bias currents supplied by R_3 and R_4 sustain the zener current independent of the signal current variations to maintain a continuous current flow through the zener diode. In essence, this continuous bias solution parallels the conversion of an amplifier output stage from class B to class A–B.

This conversion aids both the sharpness and the speed of the clamping action by not exercising the turn-on characteristic of the zener diode. First, this biased solution enhances the clamping sharpness by overriding the zener leakage current effect described in Sec. 9.1.3. As in that case, diodes D_2 and D_3 here decouple the op amp input from this leakage current, aided by the current shunting of resistors R_3 and R_4. Clamping speed improves greatly because the continuous bias

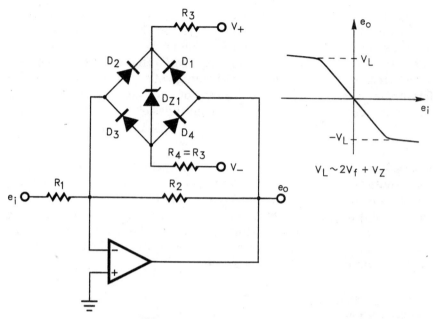

Figure 9.5 Providing continuous bias to the zener diode of Fig. 9.4 sharpens the clamp response and greatly increases clamping speed.

avoids the charging and discharging of the zener diode's capacitance. Without this bias, the preceding solution must repeatedly drive this capacitance through the full voltage range of the zener diode's current–voltage characteristic. With the bias added here, only the much smaller capacitances of the bridge diodes must be driven through their voltage extremes. Small voltage changes remain on the zener capacitance, but only those produced by signal current flow in the already biased zener diode.

To avoid turn-on delays, the R_3 resistance level should be chosen to ensure that the current through the zener diode remains above the level of the zener diode's rounded turn-on characteristic. Simple analysis guides this resistance selection to minimize the voltage change required of the zener capacitance throughout the e_o-versus-e_i response. The zener voltage still varies in response to the bridge drive of the e_o signal, which supplies current to either R_3 or R_4 through either D_4 or D_1 and the zener diode. The e_o drive can only increase the net current through the zener diode so that current reaches its minimum when $e_o = 0$. In that condition, D_1 and D_4 remain in the OFF state and $I_{Z\,min} = (V_+ - V_Z - V_-)/2R_3$. Note that the value entered for V_- in this equation should be a negative number and its presence increases $I_{Z\,min}$, in spite of the minus sign preceding the V_- term. In a secondary consideration, the selection of the zener current level potentially aids in the reduction of the thermal drift of V_L. As seen from the preceding V_L expression, that limit voltage will vary with the temperature dependencies of both $2V_f$ and V_Z. Appropriate selection of the zener current makes the temperature dependence of V_Z tend to counteract that of the $2V_f$ term.

9.2 Divider-Controlled Clamping

The preceding clamping amplifiers produce output voltage limits determined directly by zener diodes. There, the zener diodes restrict clamping voltages to specific fixed levels and preclude clamping voltages below about 1.25 V. Voltage-divider action removes these restrictions in two clamping amplifier configurations that produce virtually any clamping voltage. The first includes part of the input circuit within the zener connection span to reduce the output voltage required for forward bias of the zener diodes. However, this solution can only reduce the clamp voltage levels and potentially presents a high output resistance. The second voltage-divider alternative eliminates the zener diodes, relying upon the power-supply voltages for references and upon feedback transistors for the clamping rectification. In this way, the latter circuit permits clamping at virtually any level within the span of the power-supply voltages.

9.2.1 Voltage-divider reduction of clamping voltage

Including part of the input circuit within the zener clamping loop reduces the clamping voltage and permits fine adjustment of the clamping levels. With the preceding clamping amplifiers, the zener diodes limit the voltage developed across the R_2 feedback resistor, and this simultaneously limits the e_o output voltage at that same level. Alternately, a modification to this basic clamping approach encompasses more of the feedback network within the zener clamp loop to permit resistor adjustment and reduction of the clamping levels. As shown in Fig. 9.6, this option separates the R_1 resistor into two segments and connects the clamping zener diodes to the segments' common junction. This modification combines part of the input signal with the output signal in developing the voltage across the zener diodes. That combination reduces the output clamping voltage, and selection of the R_{1a} and R_{1b} resistors permits adjusting that voltage to values other than the fixed levels of the zener diodes.

As illustrated, this connection offers two potential output signals that supply the e_{o1} and e_{o2} terminals. The two provide clamping operations that differ in voltage range, unclamped signal gain, and output impedance. First, consider the clamping produced at the traditional e_{o1} output. In the unclamped state, input signal e_i produces a feedback

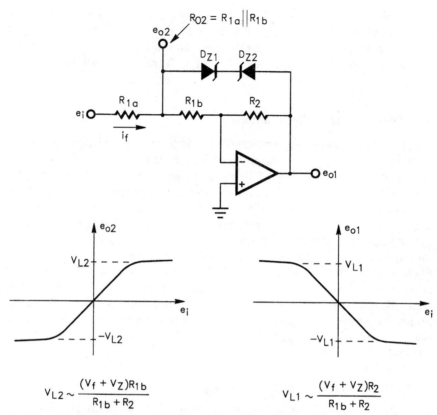

Figure 9.6 Including part of the R_1 input resistance within the clamping diodes' span reduces the clamping voltage level and permits adjustment of that level at two potential circuit outputs.

current of $i_f = e_i/(R_{1a} + R_{1b})$, and that current develops a voltage across R_2 for a signal gain of $e_{o1}/e_i = -R_2/(R_{1a} + R_{1b})$. That current also develops a voltage across the zener diodes equaling $i_f(R_{1b} + R_2)$. When that voltage reaches the turn-on level of the diodes, they divert any increase in i_f to limit the combined voltage across R_{1b} and R_2 to $V_f + V_z$. At that transition point, $i_f(R_{1b} + R_2) = V_f + V_z$, where $i_f = (V_f + V_z)/(R_{1b} + R_2)$. Then, $e_{o1} = -i_fR_2$ defines the magnitude of the clamping levels at the e_{o1} output as

$$V_{L1} \approx \frac{(V_f + V_z)R_2}{R_{1b} + R_2}$$

Thus for the e_{o1} output, R_2 and R_{1b} act as a voltage divider to reduce and adjust the clamping level from its otherwise fixed level of $V_f + V_z$.

Further evaluation of this clamping action demonstrates characteristics that distinguish it from the e_{o2} clamping performance to be described next. In the unclamped state, the e_{o1} output responds to e_i in the normal inverting amplifier mode, producing the response $e_o/e_i = -R_2/(R_{1a} + R_{1b})$. Comparing this response with the expression for V_{L1} reveals a conflict between the magnitudes of the clamping voltage and the unclamped signal gain. Both characteristics depend on the ratio of R_2 to R_{1b} precluding low clamping levels accompanied by high gain magnitudes. High gain requires that R_2/R_{1b} be large while low clamping levels require that this ratio be small. As a result, the e_{o1} output option only serves to fine-tune the clamping level for the majority of gain magnitudes accessible by the amplifier. This tuning removes the clamping level restriction otherwise imposed by limited zener voltage alternatives but does not typically offer a dramatic reduction in that level.

For greater clamp level reduction, the circuit's e_{o2} output reduces the output voltage to that developed across R_{1b}. In the unclamped state, the virtual ground of the op amp's inverting input and the i_f feedback current combine to make this voltage $e_{o2} = i_f R_{1b}$. This output signal defines the circuit's voltage gain as $e_{o2}/e_i = R_{1b}/(R_{1a} + R_{1b})$, which actually represents the attenuation of a simple voltage divider. As described, clamp turn-on results when the i_f current reaches $i_f = (V_f + V_Z)/(R_{1b} + R_2)$ and $e_{o2} = i_f R_{1b}$ defines the magnitude of the clamping levels at the e_{o2} output as

$$V_{L2} \approx \frac{(V_f + V_Z)R_{1b}}{R_{1b} + R_2}$$

Thus R_2 and R_{1b} again act as a voltage divider to reduce and adjust the clamping level from its otherwise fixed level of $V_f + V_Z$. This V_{L2} output limit differs from that of V_{L1} only in that R_{1b} rather than R_2 now serves as the output element of the divider. Expressed another way, the e_{o2} output of Fig. 9.6 reduces the clamping voltage by a factor of $(1 + R_2/R_{1b})$. This provides access to clamping levels far below the 1.25-V minimum normally imposed by zener diodes and their bandgap equivalents. It also removes the restriction to specific clamping voltage levels through the selectable gain set by R_2 and R_{1b}. Appropriate choice of these resistors sets the V_L clamping level at virtually any fraction of the voltage $V_f + V_Z$.

However, the selection of these resistors remains in compromise with two other performance characteristics of the circuit. In the unclamped state, both the voltage gain and the impedance of the e_{o2} output also depend on this selection. In that state, the e_{o2}/e_i voltage gain results from a simple voltage divider and becomes $R_{1b}/(R_{1a} + R_{1b})$, which remains less than unity. Further, the circuit output no longer benefits from the low impedance of the op amp output. Instead, the e_{o2} output terminal presents the R_{1a} resistance returning to the low impedance of the e_i source and the R_{1b} resistance returning to the virtual ground of the op amp's inverting input. Together, these two resistances define an unclamped output resistance of $R_{O2} = R_{1a} \| R_{1b}$.

Otherwise, the clamping action of Fig. 9.6 repeats characteristics of the basic clamping amplifier first presented in Sec. 9.1. The clamping levels of the e_o-versus-e_i response still display the nonzero slopes produced by the zener resistance, and the transition corners to the clamping state still follow rounded paths produced by the nonabrupt zener turn-on characteristic. In addition, mismatch between the circuit's two zener diodes produces asymmetry between the positive and negative clamping levels, as described in Sec. 9.1.4. Adaptations of this circuit employing the bridge-connected zener diode of Figs. 9.4 and 9.5 share the benefits described there.

9.2.2 Biased voltage-divider controlled clamping

The preceding clamping amplifiers all depend on zener diode voltages to define clamping levels, restricting clamping voltage alternatives to specific levels. Using a voltage-divider action, the clamping amplifier of the previous section reduces this restriction, but only for clamping voltage reduction. The preceding clamping amplifiers also produce nonzero slopes in their clamping states due to the rather high resistance characteristic of zener diodes. Replacing the zener function with power-supply bias, voltage dividers, and rectifying transistors largely removes these limitations, as shown in Fig. 9.7. There, the emitter–base junctions of transistors Q_1 and Q_2 provide the rectifying action that diverts the i_f feedback current away from R_2 to initiate clamping action. In that action, the two transistors conduct opposite-polarity i_f currents, producing bipolar limiting. Voltage dividers formed by R_3 through R_5 set the circuit's two clamping levels by biasing the bases of Q_1 and Q_2 between the e_o output signal and the two power-supply voltages. Diodes D_1 and D_2 clamp the voltage drive of these bases to protect the transistors' emitter–base junctions from reverse breakdown.

With this configuration, simple selection of the voltage dividers sets the clamping voltages at virtually any level within the range of the power supplies. Also, the lower impedances of the transistor emitter–base junctions sharpen and flatten the clamp state responses. Clamping action occurs when the e_o output voltage reaches a magnitude sufficient to forward-bias the emitter–base junction of one of the clamp transistors. Positive e_o signals turn on Q_1 when this signal drives the base potential of that transistor to $V_{B1} = V_{BE1}$, where V_{BE1} represents the forward voltage of Q_1's emitter–base junction. Note that the virtual ground at the op amp's inverting input holds the transistor emitters at

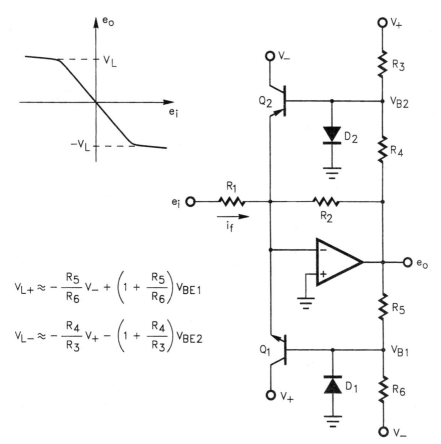

$$V_{L+} \approx -\frac{R_5}{R_6} V_- + \left(1 + \frac{R_5}{R_6}\right) V_{BE1}$$

$$V_{L-} \approx -\frac{R_4}{R_3} V_+ - \left(1 + \frac{R_4}{R_3}\right) V_{BE2}$$

Figure 9.7 Clamping amplifier where voltage dividers combine with power supplies to replace the voltage reference function of zener diodes and transistors replace the rectifying function that initiates clamping.

zero voltage, simplifying this relationship. The R_5/R_6 voltage divider supplying V_{B1} receives voltage at both ends of the divider from e_o and V_-. Superposition analysis of the two voltage effects defines $V_{B1} = (R_6 e_o + R_5 V_-)/(R_5 + R_6)$. Equating this V_{B1} expression to V_{BE1} and solving for e_o defines the magnitude of the positive clamping level as

$$V_{L+} = -\frac{R_5}{R_6} V_- + \left(1 + \frac{R_5}{R_6}\right) V_{BE1}$$

Thus when e_o reaches V_{L+}, Q_1 turns on to divert any increase in i_f away from R_2 and limit the circuit's output voltage.

Similarly, a negative e_o signal turns on Q_2 when this signal drives the base potential of that transistor to $V_{B2} = V_{BE2}$. An analysis analogous to the one that precedes defines the negative clamping level as

$$V_{L^-} = -\frac{R_4}{R_3}V_+ - \left(1 + \frac{R_4}{R_3}\right)V_{BE2}$$

Thus when e_o reaches V_{L^-}, Q_2 turns on to divert any increase in i_f away from R_2 to similarly limit the circuit's output voltage. Examination of the preceding two V_L expressions shows that this circuit permits independent setting of the positive and negative clamping voltage magnitudes to virtually any value greater than the V_{BE} of the clamping transistor but less than that of the associated power supply.

While this alternative improves the clamping action, it retains some of the limitations of previous circuits and introduces a few new ones. Power-supply noise and ripple couple through the voltage dividers to modulate V_{B1} and V_{B2} and the associated clamp voltage levels. Also, the power-supply voltage tolerances affect these levels directly. However, the noise and tolerance errors largely result from the supply's internal zener references so the end result of these two errors remains about the same as experienced with the previous zener-controlled clamping amplifiers. Still, this circuit reduces the shunt resistance of the clamp states to similarly reduce the clamp response slopes. The transistors and voltage dividers retain a residual resistance that reacts with the i_f current shunted. For positive clamp states the circuit presents a shunt resistance of $r_{e1} + (R_5 \| R_6)/\beta_1$ and for negative clamp states a resistance of $r_{e2} + (R_3 \| R_4)/\beta_2$. Here, r_e and β represent the dynamic emitter resistance and the current gain of the associated transistor. In addition, the voltage dividers of this circuit combine with parasitic capacitances to limit the clamping speed, but typically to a lesser degree than produced by the zener capacitances in the basic circuit of Fig. 9.2.

9.3   Feedback-Controlled Clamping

The clamping amplifiers of the preceding sections remove many of the performance limitations of the basic zener clamp. However, their e_o-versus-e_i transfer responses still retain rounded clamping corners and sloping clamp levels. Rounded corners result from the nonabrupt turn-on characteristics of the rectifying devices that initiate clamping. A nonzero response slope in the clamp states results from the reaction of the input signal current with the resistances of the clamping elements. Enclosing these rectifying and clamping elements within the feedback loop of an op amp reduces these nonideal characteristics by a factor equal to the circuit's loop gain. This produces nearly ideal clamping performance, as exploited by various circuits described in this and the following sections. Each provides feedback-controlled

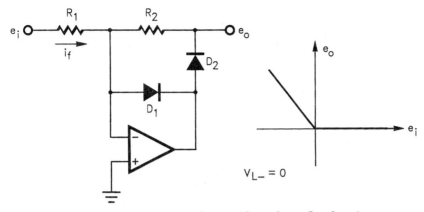

Figure 9.8 The common precision rectifier provides a sharp, flat clamping response through feedback drive that compensates for diode turn-on and resistance characteristics.

clamping with varying advantages and disadvantages and a comparison of these defines the best circuit solution for a given application.

9.3.1 Clamping response of precision rectifier

The familiar precision-rectifier connection of an op amp serves to demonstrate the basic clamping amplifier benefits provided by op amp feedback. When connected as in Fig. 9.8, this circuit produces unipolar clamping at zero voltage, permitting only positive output signals. It does so through feedback drive of the D_1 and D_2 diodes added to the feedback network of an otherwise simple inverting amplifier. In this operation, input signal e_i generates a feedback current i_f that the op amp conducts to its output through either D_1 or D_2, as directed by feedback current polarity.

Upon the application of an e_i signal, the resulting i_f current initially reacts with the very high input impedance of the op amp. The input voltage developed there drives the op amp output, through the amplifier's high open-loop gain, to divert this current through one of two feedback paths. Negative values of e_i produce negative i_f currents that can only be conducted to the op amp output through D_2, and the op amp output responds to forward-bias this diode. This condition conducts i_f through R_2 to develop an e_o output signal, making the op amp perform as a simple inverting amplifier with its familiar voltage gain of $-R_2/R_1$. Positive values of e_i produce positive i_f currents that can only be conducted to the op amp output through D_1. Then, the forward-bias of this diode diverts i_f away from R_2, disabling the inverting amplifier action. In this state, the only signal presented to R_2 and the circuit output equals that at the op amp's inverting input. With a grounded noninverting input, the op amp holds this signal voltage at zero, producing the unipolar clamping illustrated.

Feedback action sharpens and flattens this clamping response through this circuit's loop gain. Previous clamping amplifiers required direct drive of the circuit's rectifying elements by the input signal and that produced rounded clamping corners. In this case, the op amp's open-loop gain boosts that drive so that just a very small change in e_i produces the voltage change required to initiate clamping. That gain converts a small e_i change into the op amp output voltage transition required to switch the diode conduction states in Fig. 9.8, producing an extremely sharp transition. It also removes the resistance effects of the clamping elements to flatten the transfer response in the clamped state. There, any increase in e_i still produces an increase in i_f, and that current change increases the voltage developed across D_1. However, that diode's voltage no longer produces a direct influence upon the clamped output signal e_o. Feedback adjusts the op amp output voltage to absorb that diode's voltage without significantly affecting the zero-volt state maintained at the op amp's inverting input. As described, only that zero-volt signal drives the e_o output in the clamped state.

Closer evaluation of this circuit's clamping operation quantifies the actual improvement achieved. As described before, the op amp's open-loop gain boosts the signal drive of the circuit's diodes to sharpen and flatten the clamping response. However, the ability of this gain to do so remains limited by the $-R_2/R_1$ gain demanded in the linear response region. As with other feedback corrections,[5] the actual improvement achieved here depends on the loop gain or that excess remaining above the signal gain demanded. As a result, this circuit solution sharpens and flattens the transfer response by the factor $A_L = A\beta$, where A represents the op amp's open-loop gain and β represents the circuit's feedback factor in the linear state. In the transition to and in the clamped state, the op amp's gain reduces the slope of the transfer response by this factor A_L. For the circuit shown, feedback resistors R_1 and R_2 define $\beta = R_1/(R_1 + R_2)$. Other feedback-controlled clamping amplifiers described later also improve the clamping transfer response by analogous loop-gain factors.

Two practical restrictions further limit the actual performance improvement of this clamping amplifier through slew-rate limiting and output impedance. As described earlier, clamping requires that the op amp output voltage change to switch the conduction states of the circuit's two diodes. The circuit shown requires a voltage change equal to $2V_f$, where V_f represents the forward voltage drop of the diodes. Op amp slew-rate limiting defines the time required for this voltage change and, at higher frequencies, that time reintroduces rounding as well as overshoot at the clamp response corner. Also in the clamped state, this circuit presents a higher output impedance since D_2 then remains in the OFF state and the op amp output no longer controls this impedance. Instead, the virtual ground of the op amp's inverting input serves as a return for the R_2 resistor, making the clamped-state output resistance $R_O = R_2$.

9.3.2 Feedback- and divider-controlled clamping

Adding the feedback control of the precision rectifier to the divider-controlled clamping amplifier of Sec. 9.2.2 combines the benefits of the two techniques. Together, the two produce a sharp clamping response and access to very low clamping voltage levels. As shown in Fig. 9.9, this circuit combination replaces the rectifying transistors of

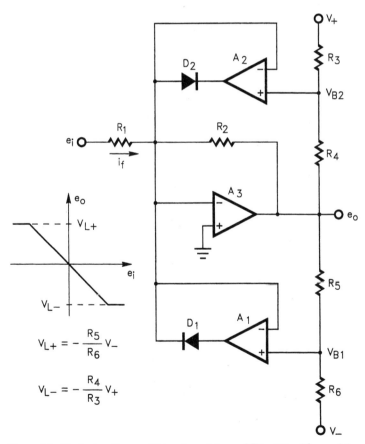

Figure 9.9 Replacing the rectifying transistors of Fig. 9.7 with precision rectifiers sharpens and flattens clamping amplifier response and also provides access to very low clamping voltage levels.

the earlier circuit with the simplified precision rectifiers formed by A_1 and A_2. Like that earlier circuit, input signal e_i generates an i_f feedback current that develops the circuit's e_o output voltage on the R_2 feedback resistor. In the unclamped state, the circuit operates as a simple inverting amplifier to produce $e_o/e_i = -R_2/R_1$. However, when the magnitude of e_o reaches a sufficient level, either the A_1 or the A_2 op amp responds to forward-bias either D_1 or D_2 and divert any increase in the i_f feedback current away from R_2. This action clamps the output voltage level at either a positive or a negative limit, producing bipolar output clamping.

A more detailed analysis quantifies the magnitudes of the two clamping levels. The A_1 and A_2 precision rectifiers of this circuit act as unity-gain voltage followers when driven to the appropriate levels by their V_{B1} and V_{B2} input voltages. These voltages drive the noninverting inputs of A_1 and A_2 and must overcome the voltage at the corresponding inverting inputs of the op amps to initiate a clamp state. Those inverting inputs connect to that of the main A_3 amplifier, where the latter maintains the zero voltage of its virtual ground input. For negative e_o signals, V_{B2} reaches this zero-volt condition when the combined effects of e_o and V_+ on the R_3/R_4 voltage divider produce $V_{B2} = 0 = e_o R_3/(R_3 + R_4) + V_+ R_4/(R_3 + R_4)$. Then, just a small negative increase in e_o drives the A_2 output negative to forward-bias D_2 and divert any increase in i_f away from R_2. Solving the last equation for e_o defines the negative clamping voltage as

$$V_{L^-} = -\frac{R_4}{R_3} V_+$$

An analogous analysis of the clamping action produced by A_1 and D_1 defines the positive clamping voltage as

$$V_{L^+} = -\frac{R_5}{R_6} V_-$$

A comparison of these clamp voltage expressions with those of the circuit in Sec. 9.2.2 directly reveals the benefits of this feedback- and divider-controlled alternative. Previously, the basic biased-voltage-divider clamp produced a positive output limit level for Fig. 9.7 of

$$V_{L^+} = -\frac{R_5}{R_6} V_- + \left(1 + \frac{R_5}{R_6}\right) V_{BE1}$$

This expression deviates from the V_{L^+} derived here in the V_{BE1} term that expresses the turn-on and resistance characteristics of the earlier transistor clamp. With the precision rectifiers here, no equivalent term results, due to the loop gain reduction of the equivalent D_1 effects. A similar comparison of the V_L- expressions for the two cases reveals the same distinction.

Here, the loop gain turns the rectifying diodes on or off in response to very small changes in e_o for much sharper and flatter clamping actions. In addition, the elimination of the V_{BE} terms of the V_L expressions provides access to much lower clamping voltage levels. As expressed by the last equation, the previous transistor-based solution required an output voltage of at least $(1 + R_5/R_6)V_{BE1}$ just to turn on the clamp. In this case, no such restriction exists, extending clamping voltage alternatives into the millivolt range.

9.3.3 Limitations of feedback- and divider-controlled clamping

While the precision rectifiers of Fig. 9.9 greatly improve clamping precision, the significances of residual errors expand and new speed and offset limitations appear. Reduced loop gain limits the actual sharpening of the clamping response, added amplifier slewing requirements compromise speed, and three op amps now contribute to clamping offsets.

First, consider the diminished loop gain produced by the circuit's added voltage dividers. As described with the precision rectifier of the preceding section, the actual degree of improvement in clamping sharpness again depends on that gain, $A_L = A\beta$. Previously, the feedback factor β only depended on the voltage divider effects of the op amp's feedback network, producing $\beta = R_1/(R_1 + R_2)$. However, in this case, the circuit adds other feedback voltage dividers to define the clamp voltage levels using R_3 through R_6. Those added dividers attenuate the feedback signal that absorbs feedback current i_f in the clamp states. As a result, these dividers reduce the degree of performance improvement by a second feedback factor.

With this circuit, the basic A_3 amplifier produces the same $\beta_3 = R_1/(R_1 + R_2)$ as before. However, for negative clamping states, the R_3/R_4 voltage divider produces the additional feedback factor $\beta_2 = R_3/(R_3 + R_4)$ through the A_2 clamping signal path. Together, these two feedback factors define the loop gain for negative clamp levels as

$$A_{L-} = A\beta_2\beta_3 = \frac{AR_1R_3}{(R_1 + R_2)(R_3 + R_4)}$$

The analogous analysis for positive clamp levels produces a controlling loop gain of

$$A_{L+} = A\beta_1\beta_3 = \frac{AR_1R_6}{(R_1 + R_2)(R_5 + R_6)}$$

As described with the preceding precision rectifier circuit, these loop gains reduce the slope of the clamping transfer response by the factor A_{L-} or A_{L+} at all points along the turn-on and clamping portions of that response.

As is often the case, the added precision here compromises circuit speed. The rectifying actions that produce clamping require that the A_1 and A_2 amplifiers slew through significant voltage ranges to initiate their clamp states. Considering the case of the A_2 precision rectifier, e_o voltage levels below the positive clamping limit make $V_{B2} < 0$ and drive the A_2 output to its negative saturation level, $-V_{O\,sat}$. When e_o reaches the level required to make $V_{B2} = 0$, the output of A_2 slews to the voltage V_f, forward biasing D_2 and initiating clamping. In this transition, the output of A_2 must slew through a voltage range equal to $V_{O\,sat} + V_f$, and the op amp's slew-rate limitation defines a finite time requirement. This delay allows the circuit output to temporarily overshoot beyond the final clamping levels, potentially a worse response error than the rounding experienced before. Previously, the basic precision rectifier of Sec. 9.3.1 reduced this delay through the inclusion of a second feedback diode. There in Fig. 9.8, the D_1 diode, connected between the op amp's inverting input and output, clamps the amplifier output to limit this output's transition voltage to $2V_f$. However, that solution cannot be applied here because the added diode would also absorb the i_f feedback currents required to develop the linear region of this circuit's e_o-versus-e_i response.

The added precision of feedback-controlled clamping also makes op amp offset voltage effects more significant. The clamping amplifiers discussed before typically depend on the virtual ground of an op amp to establish a zero voltage reference for the clamping. There, the op amp input offset voltage compromises this zero-volt condition at dc, shifting the actual clamp levels by the amount $-V_{OS}$. For those amplifiers, the offset error remained insignificant in comparison with the tolerance errors and resistance effects of the rectifying elements. Feedback-controlled clamping dramatically reduces these rectifier errors to elevate the op amp offset effects to significance. This becomes especially true when implementing the very low clamping levels made accessible with this circuit.

For the feedback- and divider-controlled clamping amplifier, all three of the A_1 through A_3 op amps contribute to offset errors. Superposition analysis considers the three amplifier effects separately to simplify the process and demonstrate the relative effect of each amplifier. In Fig. 9.10, the V_{OS1} through V_{OS3} input offset voltages of the three op amps compromise the ideal zero-volt virtual-ground condition underlying the ideal analysis of the preceding section. Using superposition, first consider the $V_{OS1} = V_{OS2} = 0$ condition to define the effect of V_{OS3}. The latter offset again shifts the ideal virtual ground and the resulting clamp levels by the

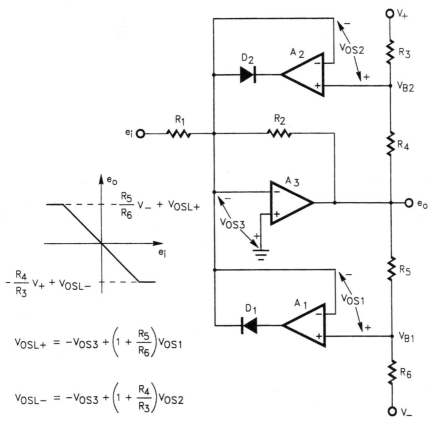

Figure 9.10 Input offset voltages of three op amps combine to produce offsets in the positive and negative clamping levels of Fig. 9.9.

amount $-V_{OS3}$. Next considering $V_{OS2} = V_{OS3} = 0$, the V_{OS1} offset shifts the switching point of the positive clamping level from the $V_{B1} = 0$ condition to $V_{B1} = V_{OS1}$. The R_5/R_6 voltage divider supplying V_{B1} attenuates the e_o influence on V_{B1}, amplifying the V_{OS1} effect upon the clamping level to $(1 + R_5/R_6)V_{OS1}$. Similarly, the $V_{OS1} = V_{OS3} = 0$ condition defines the corresponding offset shift in the negative clamping level as $(1 + R_4/R_3)V_{OS1}$. Combining the offset effects of the circuit's three op amps defines the associated shifts in the positive and negative clamping levels as

$$V_{OSL+} = -V_{OS3} + \left(1 + \frac{R_5}{R_6}\right)V_{OS1}$$

and

$$V_{OSL-} = -V_{OS3} + \left(1 + \frac{R_4}{R_3}\right)V_{OS2}$$

9.4   Voltage-Controlled Clamping

Reference voltage control of clamping levels also accesses clamping levels at virtually any voltage within the circuit's operating range. In addition, this voltage control permits electronic variation of clamping levels for test and amplitude modulation requirements. Combining voltage control with the feedback control described in Sec. 9.3 retains the clamping precision described there. For this combined purpose, adaptations of the earlier precision rectifier clamp produce precise and versatile unipolar and bipolar clamping amplifiers. For these amplifiers, the precision rectifier serves as a switch that enables or disables signal coupling to the circuit output and signal summation places the switching action under the control of an applied reference voltage.

9.4.1   Unipolar voltage-controlled feedback clamping

The basic precision rectifier described in Sec. 9.3.1 produces unipolar clamping, but only for a zero-volt clamp level. In addition, that circuit constrains the clamping transition to the origin of the e_o-versus-e_i response. General clamping applications require offsetting this transition along both the e_o and the e_i axes to develop a clamping action at nonzero levels of both signals. Adding a reference voltage and signal summation to the basic precision rectifier performs these offsetting functions, as illustrated in Fig. 9.11. There, the reference voltage V_R, the circuit's two R_3 resistors, and the amplifier formed with A_2 introduce signal summation at the input and the output of the precision rectifier formed with A_1. Summation at the input shifts the clamping transition along the e_i response axis and the output summation shifts it along the e_o axis. The A_2 summing amplifier also buffers the final circuit output from the higher output resistance presented by the precision rectifier in the clamp state. Note that resistors bearing the same numerical designator, such as R_2, indicate equal-valued resistances. The three R_2 resistors shown must be of the same value, as must the two R_3 resistors, to assure a critical gain balance. Lack of this balance introduces major output offset errors, as will be described.

Analysis clarifies this gain balance requirement and defines the circuit's transfer response parameters. As shown, the e_{o2}-versus-e_i response displays linear and clamped regions representing two circuit states. Circuit operation switches from the linear state to the clamped state when the polarity of feedback current i_f reverses. That current's polarity determines whether D_1 or D_2 conducts to control the instantaneous circuit function of the A_1 op amp. Input signal e_i varies this current as described by the relationship $i_f = e_i/R_1 + V_R/R_3$. Polarity reversal for i_f and the resulting diode switching occur when i_f passes through zero, where $e_i = -(R_1/R_3)V_R$. For $e_i < -(R_1/R_3)V_R$, $i_f < 0$ to

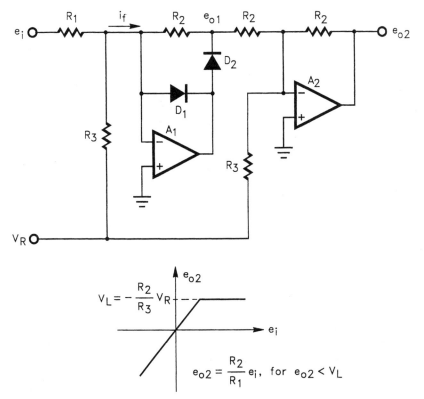

Figure 9.11 Adding signal summation at the input and output of a precision rectifier permits controlling the clamping transition point through reference voltage V_R.

forward-bias D_2 and reverse-bias D_1, producing a circuit state modeled by Fig. 9.12. There, a short circuit replaces the forward-biased D_2 and an open circuit replaces the reverse-biased D_1 to simplify the analysis. In practice, the D_2 voltage drop of this condition shifts the output voltage of A_1 but does not directly influence the input-to-output signal path through the circuit's resistors. As described for the basic precision rectifier, the high loop gain of the A_1 circuit virtually eliminates the signal influence of this diode voltage.

As modeled in Fig. 9.12, the clamping amplifier appears as two series-connected summing amplifiers. The first of these produces $e_{o1} = -(R_2/R_1)e_i - (R_2/R_3)V_R$, which displays the desired linear e_{o1}-versus-e_i relationship, having a gain of $-R_2/R_1$. This e_{o1} expression also displays the undesired offset term $-(R_2/R_3)V_R$. However, as will be seen, that offset must be accepted here to permit later control of the clamped state. For this linear state, the second summing amplifier of the model removes this offset to produce the final output signal $e_{o2} = -e_{o1} - (R_2/R_3)V_R$, or

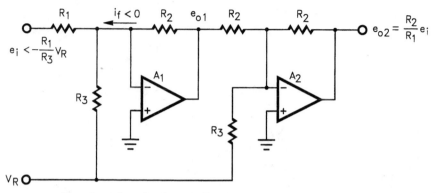

Figure 9.12 Negative values for the i_f feedback current configure the circuit of Fig. 9.11 as two series-connected summing amplifiers for a linear signal response of $e_{o2} = (R_2/R_1)e_i$.

$$e_{o2} = \frac{R_2}{R_1}\,e_i, \qquad \text{for } e_{o2} < V_L$$

where V_L represents the clamping level defined next.

As mentioned before, the offset removal of the linear state requires establishing a gain balance through resistor selection. The equal-valued R_2 resistors combined with the equal-valued R_3 resistors assure this condition for the case illustrated. Other resistor combinations will also remove the offset error, in accessing other signal and clamping gain options, but the selection of those combinations should still be guided by the offset removal requirement. In any case, mismatches in the circuit's resistors typically produce a residual error that dominates offset performance in this linear circuit state. Improved resistor matching and lower V_R values reduce this offset error, but only to the degree permitted by the input offset voltages of the circuit's two op amps. Including those offsets in Fig. 9.12, setting $e_i = 0$ and solving for e_{o2} defines the minimum output offset error as

$$V_{OSO\,min} = \left(1 + \frac{R_2}{R_1} + \frac{R_2}{R_3}\right)V_{OS1} - \left(2 + \frac{R_2}{R_3}\right)V_{OS2}$$

where V_{OS1} and V_{OS2} represent the input offset voltages of the two op amps.

Next consider the clamped state as developed by $e_i > -(R_1/R_3)V_R$. This produces $i_f > 0$ to forward-bias D_1 and reverse-bias D_2, producing the circuit state modeled by Fig. 9.13. There, the short-circuit feedback of the A_1 op amp presents this amplifier's virtual ground within the signal input circuit, maintaining a zero-volt condition that shunts away

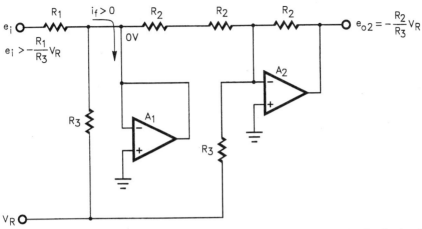

Figure 9.13 Positive values for the i_f feedback current of Fig. 9.11 disable the A_1 signal path, relegating control of the e_{o2} output to V_R and producing a clamped voltage level of $-(R_2/R_3)V_R$.

the i_f feedback current. This shunting removes the influence of e_i upon the e_{o2} output signal to develop the clamped state. As modeled, the control of the final e_{o2} output signal then depends on a summing amplifier formed with A_2 and driven from two summing inputs. One input couples through two series-connected R_2 resistors from a zero-volt source and produces no e_{o2} output. The other input couples from reference voltage V_R and defines the clamped output voltage as e_{o2} reaches

$$V_L = -\frac{R_2}{R_3} V_R$$

Resistor tolerance errors and op amp offset voltages produce deviations from this ideal clamp level. The V_L equation directly reflects the effects of resistance errors, which can be controlled through resistor tolerance selection. However, even with perfect resistors, the input offset voltages of the circuit's two op amps produce a residual output offset. Including those input offsets in Fig. 9.13 and considering $e_i = 0$ shows that these op amp offsets produce an output clamping error of

$$V_{OSL} = \frac{V_{OS1}}{2} - \left(1.5 + \frac{R_2}{R_3}\right)V_{OS2}$$

where V_{OS1} and V_{OS2} represent the input offset voltages of the two op amps.

Comparing the preceding V_L result with the e_{o2} result of the circuit's linear state reveals independent voltage gains defining the lin-

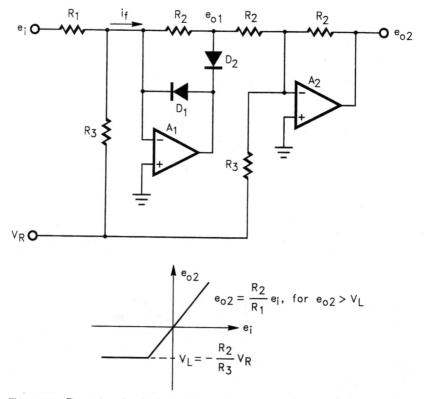

Figure 9.14 Reversing the diode directions of the clamping amplifier of Fig. 9.11 produces a lower rather than upper bound for the e_{o2} response.

ear and clamped circuit states. From before, the circuit produces either a linear output of $e_{o2} = (R_2/R_1)e_i$ or a clamped output of $V_L = -(R_2/R_3)V_R$. For a given value of R_2, selecting R_1 defines the (R_2/R_1) linear gain that amplifies e_i and selecting R_3 defines the $-R_2/R_3$ clamping gain that amplifies V_R. Note that the inversion or minus sign of the latter gain requires a negative V_R voltage to produce the positive V_L clamping bound illustrated in Fig. 9.11. However, positive V_R voltages will still control clamping action but shift the e_{o2}-versus-e_i response of that figure below the e_i axis.

Reversing the directions of the circuit's two diodes produces an analogous circuit with a lower rather than upper bound, as illustrated in Fig. 9.14. Analysis of this alternative directly parallels the preceding case with only the reversal of the < and > signs in the text and in the analysis models of Figs. 9.12 and 9.13. For Fig. 9.14, that analysis produces

$$e_{o2} = \frac{R_2}{R_1} e_i, \quad \text{for } e_{o2} > V_L$$

and repeats the earlier results of

$$V_L = -\frac{R_2}{R_3} V_R$$

and

$$V_{OSL} = \frac{V_{OS1}}{2} - \left(1.5 + \frac{R_2}{R_3}\right) V_{OS2}$$

9.4.2 Bipolar voltage-controlled feedback clamping

Adding a second precision rectifier to the preceding circuit develops bipolar clamping action for voltage control of both upper and lower signal bounds.[6] The resulting clamping action remains very similar to that described in greater detail for the unipolar clamping amplifier of the preceding section. The bipolar alternative, shown in Fig. 9.15, employs two series-connected precision rectifiers followed by a summing amplifier that also serves to buffer the circuit output. As before, the rectifiers function as switches that either enable or disable signal flow from the e_i terminal through the $R_1 - R_2$ resistor string to the e_{o3} output. In this path, the summation of control voltage V_R through the R_3 resistances offsets the switching transitions to define the clamping levels.

Two cautions should be observed when implementing this circuit. The control voltage V_R must be a negative voltage to produce nonoverlapping clamping levels that would otherwise produce total signal blanking. Analyses that follow define input and output clamp level limits that clarify this requirement. Also, resistors bearing the same designator, such as R_2, again indicate equal-valued resistances. Similarly, the $R_3/2$ designator indicates a resistance of one-half the value of R_3. Resistance deviations from those so indicated introduce major output offset errors in both the linear and the clamped states.

Analysis clarifies these offset effects and defines the circuit's transfer response parameters. As illustrated, the e_{o3}-versus-e_i response displays one linear and two clamped regions representing three circuit states. Like before, circuit operation switches from the linear state to a clamped state when the polarity of the i_f feedback current reverses for either A_1 or A_2. For both of these amplifiers, that current's polarity determines whether the corresponding D_1 or D_2 conducts to control the instantaneous circuit function of the associated op amp. Input signal e_i

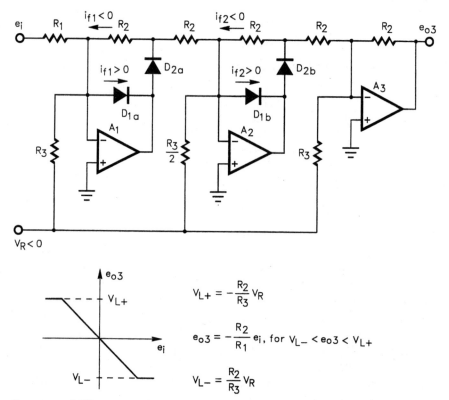

Figure 9.15 Adding a second precision rectifier to the clamping amplifier of Fig. 9.11 produces bipolar limiting.

varies the i_{f1} and i_{f2} currents to drive the e_{o3} output through the linear region to either of the circuit's two clamp states.

First, consider the linear response region where negative feedback currents for both A_1 and A_2 configure the circuit as modeled in Fig. 9.16. A comparison of this illustration with Fig. 9.15 reveals that the $i_{f1} < 0$ and $i_{f2} < 0$ shown here forward-bias D_{2a} and D_{2b} and reverse-bias D_{1a} and D_{1b} of the earlier figure. To simplify the analysis, the model here replaces the forward-biased diodes with short-circuit connections and the reverse-biased ones with open circuits. This configures the circuit model as three series-connected summing amplifiers that define the voltage gain of the linear region and illustrate a first resistor matching requirement. The first amplifier of this series produces $e_{o1} = -(R_2/R_1)e_i - (R_2/R_3)V_R$, which displays the desired e_{o1}-versus-e_i relationship of a linear amplifier. This e_{o1} expression also displays an undesired offset term, $-(R_2/R_3)V_R$, as required for clamp level voltage

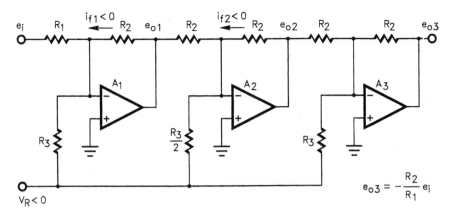

Figure 9.16 Negative i_{f1} and i_{f2} feedback currents for A_1 and A_2, as controlled by e_i and V_R, configure the clamping amplifier of Fig. 9.15 as three series-connected summing amplifiers to produce a linear response.

control, but later removed through subsequent summations. The second amplifier of the series produces $e_{o2} = -e_{o1} - 2(R_2/R_3)V_R = (R_2/R_1)e_i - (R_2/R_3)V_R$, expressing similar gain and offset conditions. Finally, the third amplifier produces $e_{o3} = -e_{o2} - (R_2/R_3)V_R = -(R_2/R_1)e_i$ to remove the offset term and define the response of the linear range as

$$e_{o3} = -\frac{R_2}{R_1}e_i, \qquad \text{for } V_{L-} < e_{o2} < V_{L+}$$

where V_{L+} and V_{L-} represent the clamping limits to be defined.

As mentioned before, the offset removal of the linear state requires establishing a gain balance through resistor selection. The equal-valued R_2 resistances combined with the R_3 resistance ratio shown assure this condition for the case illustrated. Other resistor combinations also remove the offset error in accessing alternate clamping options such as different clamping magnitudes for the positive and negative bounds. However, the selection of those combinations should still be guided by the offset removal requirement. In any case, mismatches in the circuit's resistors typically produce a residual error that dominates offset performance in this linear circuit state. Improved resistor matching and lower V_R values reduce this offset error, but only to the degree permitted by the input offset voltages of the circuit's three op amps. Including those offsets in Fig. 9.16, setting $e_i = 0$, and solving for e_{o3} defines the minimum output offset error as

$$V_{OSO\,min} = -\left(1 + \frac{R_2}{R_1} + \frac{R_2}{R_3}\right)V_{OS1} + 2\left(1 + \frac{R_2}{R_3}\right)V_{OS2} - \left(2 + \frac{R_2}{R_3}\right)V_{OS3}$$

where V_{OS1}, V_{OS2}, and V_{OS3} represent the input offset voltages of the circuit's op amps.

9.4.3 Voltage-controlled bipolar clamping action

As described, the linear response state of this bipolar clamping amplifier requires that $i_{f1} < 0$ and $i_{f2} < 0$ for the A_1 and A_2 op amps. A polarity reversal of either of these feedback currents initiates a clamp state, as seen from Fig. 9.15, where a feedback current flowing into the inverting inputs of A_1 or A_2 forward-biases either D_{1a} or D_{1b} and reverse-biases the corresponding D_{2a} or D_{2b}. First, consider the clamp state affected by A_1 when i_{f1} makes the transition from negative to positive, as modeled in Fig. 9.17. This illustration again replaces forward- and reverse-biased diodes with short- and open-circuit connections. From this figure, $i_{f1} = e_i/R_1 + V_R/R_3$ and the associated polarity reversal occurs when i_{f1} passes through zero, for which $e_i = -(R_1/R_3)V_R$.

Input signals having a magnitude of $e_i > -(R_1/R_3)V_R$ maintain the condition illustrated and analysis quantifies the resulting negative output clamp state. In Fig. 9.17, the short-circuit feedback of the A_1 op amp presents this amplifier's virtual ground within the input signal path, maintaining a zero-volt condition that shunts away the i_{f1} feedback current. This shunting removes the influence of e_i upon the e_{o3} output signal to develop a clamped state. In this state, the control

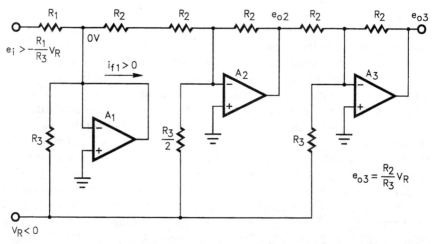

Figure 9.17 A positive i_{f1} feedback current for A_1 introduces a virtual ground in the e_i-to-e_{o3} signal path to establish the circuit's negative output clamp state.

of the final e_{o3} output signal depends on the two summing amplifiers formed with A_2 and A_3. The first of these develops $e_{o2} = -2(R_2/R_3)V_R$ and the second develops $e_{o3} = -e_{o2} - (R_2/R_3)V_R$. Combining these last two equations defines the circuit's negative clamp level as

$$V_{L-} = \frac{R_2}{R_3} V_R$$

where V_R represents a negative voltage.

Resistance tolerance errors and op amp offset voltages produce deviations from this ideal clamp level, as defined by straightforward analyses. Direct inclusion of the resistor errors in this V_{L-} equation defines the associated offset error produced by the R_2 and R_3 resistances in conjunction with the V_R bias. Improved resistor matching and lower V_R magnitudes reduce this offset error, but only to the degree permitted by the input offset voltages of the circuit's three op amps. Including those offsets in Fig. 9.17, setting $e_i = 0$, and solving for e_{o3} defines the associated minimum output offset error as

$$V_{OSL-min} = -\frac{V_{OS1}}{2} + \left(1.5 + 2\frac{R_2}{R_3}\right)V_{OS2} - \left(2 + \frac{R_2}{R_3}\right)V_{OS3}$$

where V_{OS1}, V_{OS2}, and V_{OS3} represent the input offset voltages of the op amps.

Next, consider the clamp state affected by A_2 when its i_{f2} feedback current makes the polarity transition, as illustrated in Fig. 9.18.

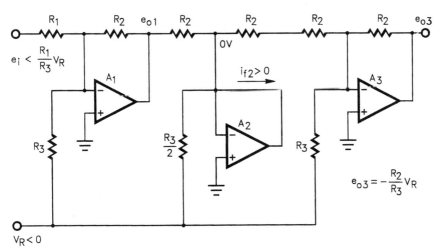

Figure 9.18 A positive i_{f2} feedback current for A_2 introduces a similar virtual ground in the e_i-to-e_{o3} signal path to establish the circuit's positive output clamp state.

There, $i_{f2} = e_{o1}/R_2 + 2V_R/R_3$, and the polarity reversal occurs when e_{o1} reaches $e_{o1} = -2(R_2/R_3)V_R$. At this transition, the A_1 function remains in its linear mode, continuing the $e_{o1} = -(R_2/R_1)e_i - (R_2/R_3)V_R$ response. Combining the last two equations defines the associated limit transition of A_2 at $e_i = (R_1/R_3)V_R$. As modeled here, the short-circuit feedback of the A_2 op amp presents this amplifier's virtual ground within the signal path to shunt away the i_{f2} feedback current. This shunting again removes the influence of e_i upon the e_{o3} output signal to develop a second clamped state. In this state, the control of the final e_{o3} output signal only depends on a summing amplifier, formed with A_3 and driven from two summing inputs. One input couples through two series-connected R_2 resistors from a zero-volt source and produces no e_{o3} output. The other input couples from the reference voltage V_R and defines the clamped output voltage as an e_{o3} level of

$$V_{L+} = -\frac{R_2}{R_3} V_R$$

where V_R represents a negative voltage.

Resistance tolerance errors and op amp offset voltages again produce deviations from this ideal clamp level, as defined by straightforward analyses. Direct inclusion of the resistor errors in this V_{L+} equation defines the associated offset error produced by R_2 and R_3 resistances in conjunction with the V_R bias. Improved resistor matching and lower V_R magnitudes reduce this offset error, but only to the degree permitted by the input offset voltages of the circuit's three op amps. Including those offsets in Fig. 9.18, setting $e_i = 0$, and solving for e_{o3}, defines the minimum output offset error as

$$V_{OSL+min} = \frac{V_{OS2}}{2} - \left(1.5 + \frac{R_2}{R_3}\right)V_{OS3}$$

where V_{OS1}, V_{OS2}, and V_{OS3} represent the input offset voltages of the op amps.

Comparing the preceding V_{L-} and V_{L+} results with that of the circuit's linear state reveals independent voltage gain magnitudes for the linear and clamped circuit states. From before, the circuit produces either a linear output of $e_{o3} = -(R_2/R_1)e_i$ or a clamped output of $V_{L-} = (R_2/R_3)V_R$ or $V_{L+} = -(R_2/R_3)V_R$. For a given value of R_2, selecting R_1 defines the $-R_2/R_1$ gain that amplifies e_i in the linear region. Then, selecting R_3 defines the R_2/R_3 gain magnitude that amplifies V_R in both the positive and the negative clamp states. Note again that this circuit requires a negative V_R value to produce nonoverlapping clamping bounds that would otherwise produce total signal blanking.

The resulting e_{o3}-versus-e_i response of Fig. 9.15 illustrates the circuit action with the upper and lower bounds produced by a negative V_R reference voltage.

References

1. J. Graeme, "Clamping Circuits Improve Precision of Bipolar Limiters," *EDN*, p. 189, June 22, 1989.
2. J. Graeme, *Designing with Operational Amplifiers; Applications Alternatives*, McGraw-Hill, New York, 1977.
3. J. Graeme, *Applications of Operational Amplifiers*, McGraw-Hill, New York, 1973.
4. R. Pease, "Bounding, Clamping Techniques Improve Circuit Performance," *EDN*, p. 277, November 10, 1983.
5. J. Graeme, *Optimizing Op Amp Performance*, McGraw-Hill, New York, 1997.
6. Graeme, op cit., 1977.

10

Phantom Amplifiers

For years the venerable phantom circuit reduced wiring complexity in remote instrumentation[1] but failed to serve lower-frequency signals. Phantom amplifiers continue the wiring reduction benefit and extend the technique down to dc. These amplifiers instrument the differential measurement of remote signals that inherently acquire common-mode errors and potentially develop crosstalk through shared common returns. As described with the instrumentation amplifiers of Chap. 7, differential measurement greatly improves remote monitoring accuracy by rejecting the effects of coupled noise and ground potential differences. As described here, differential measurement also removes the crosstalk effects of remote multichannel instrumentation.

However, this basic differential instrumentation requires two connecting wires for each signal monitored. In multichannel instrumentation systems, phantom circuits and phantom amplifiers reduce these wiring requirements by one-third. They do so by coupling a third differential signal as the difference between the common-mode signals of two other monitor channels. Differential measurements with the transformer-coupled phantom circuit become impractical at lower frequencies because of the physical size of the transformers then required. Replacing the transformers with differential amplifiers forms phantom amplifiers that provide the same wiring efficiency but without the low-frequency limitation.

10.1 Remote Differential Monitoring

Several sources introduce transmission errors in remote monitoring, including noise pickup, ground-potential differences, and crosstalk. When applied correctly, differential measurement largely removes all three of these errors. For a single-channel application, differential measurement on a two-wire pair removes the errors of noise pickup and ground-potential differences. For multichannel applications, the temptation to use a single ground return line introduces crosstalk through line voltage drops developed with parasitic line impedances. At first inspection, avoiding this crosstalk seems to require a separate two-wire pair for each signal monitored. Later evaluation of phantom circuits and phantom amplifiers reveals the means to couple a third signal through the two two-wire pairs dedicated to other signals.

10.1.1 Rejecting noise and ground-potential errors

Simple differential measurement removes the first two error sources of remote monitoring, as illustrated in Fig. 10.1. There, Fig. 10.1a shows the most basic signal monitor configuration in which output

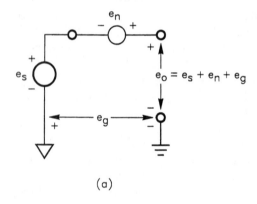

(a)

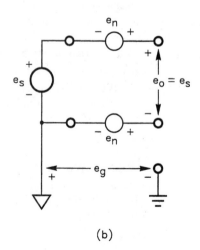

(b)

Figure 10.1 Noise and a ground-potential difference introduce errors in single-ended signal monitoring (a), but differential monitoring removes those errors (b).

signal e_o ideally tracks input signal e_s. However, as modeled, the two other signals e_n and e_g interfere with this measurement due to the large distance separating the signal source and the monitored output. Just the distance itself results in a difference in ground potentials to produce error signal e_g. Added to this, spurious coupling introduces the noise signal e_n, picked up from the environment by the long wire that spans the separation. In this basic case, both e_g and e_n combine with e_s to degrade the measured signal e_o and produce

$$e_o = e_s + e_n + e_g$$

Switching to differential measurement in Fig. 10.1b removes both of these error sources. The ground-potential difference e_g no longer influences the output because e_s and e_o now share the same common, just through different return connections. Further, the differential measurement rejects the e_n voltage drops through the inherent common-mode rejection of that measurement. There, the two wires of Fig. 10.1b experience the same noise environment and pick up the same noise signal e_n. However, the differential measurement of e_o detects only the difference between the signals on the two wires and the two e_n noise effects cancel, making $e_o = e_s$.

10.1.2 Avoiding crosstalk

Even with differential sensing, remote monitoring remains vulnerable to line voltage drops that potentially introduce crosstalk. In such applications, the temptation to utilize one ground return line for multiple signal transmissions frequently produces crosstalk errors, as modeled in Fig. 10.2. There, differential sensing connections monitor two signals to effectively remove the effects of ground potential differences and noise pickup, as described before. For simplicity, the figure here excludes those errors to focus on the crosstalk errors produced by line resistances and inductances. Those parasitic impedances produce line voltage drops in conjunction with the load currents supplied to the R_{L1} and R_{L2} output loads shown. Through their connecting lines, input signals e_{s1} and e_{s2} drive R_{L1} and R_{L2} to produce load currents i_{L1} and i_{L2}. These currents react with the line impedances, producing the line voltage drops e_{L1} and e_{L2}. In remote monitoring, these line drops produce significant errors due to the higher line impedances encountered. For example, 22-gage copper wire introduces 16 Ω of parasitic resistance per 1000 ft, and hundreds of ohms result with remote wire runs. Similarly, wire of almost any gage introduces around 180 μH of parasitic inductance[2] per 1000 ft.

As modeled here, the resulting line drops couple signals between the two monitor loops to produce crosstalk. This coupling results from the

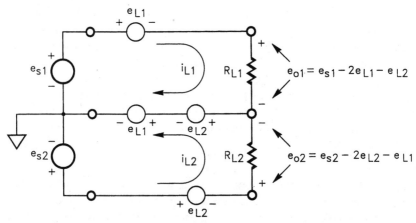

Figure 10.2 A ground return line common to two differential signals introduces crosstalk through line voltage drops developed by load current reactions with parasitic line impedances.

e_{L1} and e_{L2} line voltage drops developed in the common return line shared by the two loops. Because of these error signals, the e_{o1} and e_{o2} output signals include signal components developed by both input signal sources. Examination of the e_{o1} output defines its signal as

$$e_{o1} = e_{s1} - 2e_{L1} - e_{L2}$$

Here, the term e_{L2} results from a line drop developed by e_{s2} and coupled into the e_{s1} monitor loop by the shared common return. That intermixing of e_{s1} and e_{s2} signals represents crosstalk. An analogous crosstalk term occurs in the e_{o2} output due to the e_{L1} drop introduced to the common line by e_{s1}. These e_{o1} and e_{o2} output signals also contain error terms developed by the intended signal sources. As expressed before, the e_{o1} output contains a second error term of $-2e_{L1}$, and the e_{o2} output develops an analogous $-2e_{L2}$ term. However, these error terms represent only gain errors, which remain far less serious than the crosstalk components. Subsequent gain adjustments remove the gain error effects but crosstalk signals, once introduced, defy such simple correction.

When removing crosstalk, avoidance offers the best solution, as illustrated in Fig. 10.3. There, separate common return lines serve each signal loop to isolate the effects of their e_{L1} and e_{L2} line drops. With this connection, each monitor loop contains only the line drops produced by that loop's signal, and this avoids the crosstalk described before. Then, the e_{o1} output delivers

$$e_{o1} = e_{s1} - 2e_{L1}$$

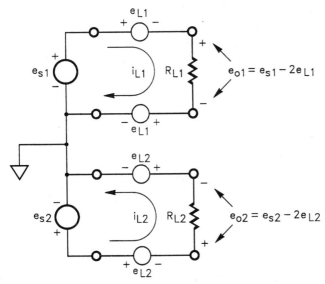

Figure 10.3 Providing separate ground returns for each signal monitored removes the crosstalk of Fig. 10.2 but retains the gain error produced by line voltage drops.

without the e_{L2} crosstalk term of the previous case. The e_{o2} output delivers an analogous $e_{o2} = e_{s2} - 2e_{L2}$. These two output signals continue to develop correlated gain error signals of $2e_{L1}$ or $2e_{L2}$, but they avoid the uncorrelated error signals that represent crosstalk. Thus a separate pair of wires for each signal assures accurate differential monitoring of remote signals. However, this requirement makes mundane wire a major cost factor in the implementation and maintenance of remote monitoring systems.

10.2 Phantom Circuits

The seemingly inevitable wiring requirement described in the preceding section receives dramatic relief through a configuration known as a phantom circuit.[3] This venerable circuit instruments differential measurement in a manner that reduces the wiring requirements of multichannel systems by one-third. Conceptually, the phantom approach transmits a third differential signal through common-mode signals introduced on two other wire pairs. There, each wire pair transmits its own differential signal, which remains immune to any common-mode signals, but intentionally introducing different common-mode signals on these two pairs transmits a third differential signal without the need for a third set of wires. This third signal transmits as the difference between the two common-mode signals, as though a third or phantom wire pair were present. Historically, transformers implemented the phantom transmission through their ability to couple both differential and common-mode signals. Examination of the transformer case here defines the fundamental circuit requirements that later permit replacement of the transformers with differential amplifiers.

10.2.1 Common-mode signal transmission

Before examining the complete phantom circuit, consider how transformers permit the introduction and then the separation of a common-mode signal in a differential signal transmission. These capabilities underlie the phantom circuit function described later. In Fig. 10.4, an input transformer begins the transmission process by removing any common-mode signal present in the input signals e_1 and e_2. The transformer's differential output responds only to the difference between these signals, e_s. Then with the 1:1 transformer ratio shown, signal e_s transfers directly to the transformer's differential output, where that signal contains no inherent common-mode component and remains free for common-mode control.

The signal drive of the transformer's output center tap produces this common-mode control through e_3. This drive shifts both transformer output terminals by the same voltage and does not alter the transformer's differential output signal. Signal e_s simply becomes referenced to e_3, rather than ground, and the voltages of the two transmission lines swing about this e_3 reference level. With this drive combination, the single-ended signals at the transformer's two output terminals become $e_3 + e_s/2$ and $e_3 - e_s/2$, preserving e_s as a differential output signal while conducting signal e_3 as a common-mode signal.

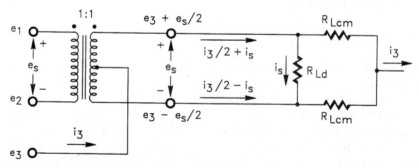

Figure 10.4 To extend wiring efficiency, signal e_3 drives the center tap of an input transformer coupling differential signal e_s, making e_3 part of the total transmission of a two-wire pair.

In practice, the associated load currents that flow in finite line impedances potentially mix the differential and common-mode signals. To prevent this crosstalk, the transformer circuit shown requires balanced impedances for the common-mode signal on the two output lines. Then, a common-mode signal driving two matched lines develops equal line currents and equal line voltage drops that the differential output measurement rejects. The illustrated configuration assures this balanced impedance condition through three requirements. Starting at the transformer's secondary, the center-tap connection of the common-mode signal assures equal winding impedances between the tap and the two output lines. Next, matched lengths for the two output lines introduce equal parasitic impedances. Finally, output loading with two equal R_{Lcm} common-mode loads completes the impedance balance. Any impedance driven by common-mode current i_3 beyond the R_{Lcm} loads remains common to the two lines and introduces no differential imbalance.

Given the balanced conditions described, the common-mode load current i_3 faces equal impedances along the circuit's two lines, making this current divide equally between them. The $i_3/2$ common-mode currents do not alter the differential signal voltage reaching the differential load R_{Ld}. Only the load current i_s from differential signal e_s flows in R_{Ld}, and the i_3 common-mode currents, drawn by the balanced common-mode loads, bypass this differential load. Further, differential measurement of the voltage across R_{Ld} rejects the matching line voltage drops developed by the $i_3/2$ currents, just as the example in Fig. 10.1b rejected the two equal e_n noise signals. This avoids crosstalk from the common-mode signal into the differential output to the degree permitted by impedance balance and transformer center tap accuracy.

10.2.2 Common-mode signal separation

The common-mode drive of Fig. 10.4 introduces an additional signal to a wire pair at the input side of a monitor connection. At the output side, the common-mode and differential signals must be separated. For this purpose, a transformer center tap again provides the solution, as seen by considering the reverse operation of the transformer in Fig. 10.4. Switching the input and output roles of the transformer provides the required circuit balance, as demonstrated in Fig. 10.5. There, the transmitted $e_3 + e_s/2$ and $e_3 - e_s/2$ signals drive the inputs of a receiving transformer. This signal drive produces a differential input signal equal to e_s, which transfers directly to this transformer's output. Simultaneously, the e_3 common-mode signal of the two lines drives the transformer's center tap to transfer that signal separately to another circuit output. In practice, line voltage drops reduce both the differential and the common-mode signals transferred, but under

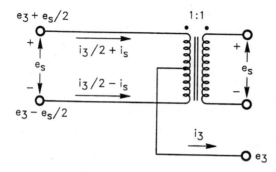

Figure 10.5 At the monitor system output, a center-tapped transformer separates the e_3 common-mode signal from its differential e_s counterpart.

the balanced conditions discussed before, those drops introduce only gain errors and not crosstalk. For simplicity in this discussion, this illustration excludes those line voltage drops.

Crosstalk from the differential signal e_s into the common-mode signal e_3 only results from an asymmetry between the two voltage signals of the transmission. Ideally, the center tap of the receiving transformer follows the midpoint or average of the voltages on the two transmission lines. There, the ideal $e_3 + e_s/2$ and $e_3 - e_s/2$ signals shown make this midpoint voltage equal the common-mode signal e_3 with the $e_s/2$ components canceling. Three error sources potentially disturb this cancellation to produce crosstalk. First, a center tap error in the transmitting transformer can shift more of the e_s signal to one line than the other, disturbing the $\pm e_s/2$ symmetry expressed before. Second, mismatched line voltage drops produce a similar effect. Finally, the center tap error of the receiving transformer disturbs the cancellation of even perfect $\pm e_s/2$ transmitted signal components. Any of these three effects adds a portion of e_s to the e_3 output signal.

10.2.3 Practical phantom circuits

Combining these two transformer connections, the generalized phantom circuit of Fig. 10.6 provides differential transmission of three signals using just two wire pairs. The initial discussion of this circuit neglects all common and ground connections, considering only differential signals. Later, the conversion of the basic phantom circuit to more generalized instrumentation reintroduces those ground connections. In Fig. 10.6, the differential input signals $e_{s1} = e_1 - e_2$ and $e_{s3} = e_5 - e_6$ drive two pairs of wires through transformers T_1 and T_3, and these transformers remove any input common-mode signals accompanying e_{s1} and e_{s3}. Similarly at the circuit output, transformers T_2 and T_4 separate those differential signals from any common-mode signals

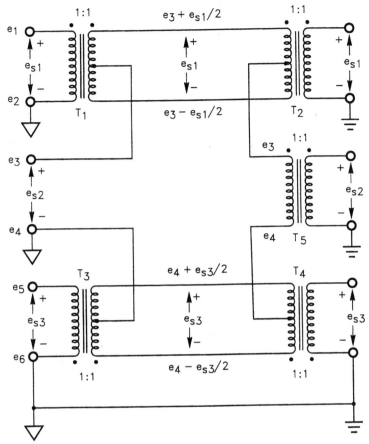

Figure 10.6 Combined transmission of differential and common-mode signals adds a third or phantom transmission to the differential coupling of two pairs of wires.

introduced on the lines. Thus signals e_{s1} and e_{s3} couple from input to output on their dedicated wire pairs just as though an ideal transmission environment existed and the transformers were not present.

However, the introduction of the transformers provides a free ride for the third differential signal, $e_{s2} = e_3 - e_4$. This signal transmits as the difference between the two common-mode signals introduced on the wire pairs dedicated to e_{s1} and e_{s3}. Through a center-tap connection, signal e_3 couples from input to output as a common-mode signal on the wire pair carrying differential signal e_{s1}. On the output side of the circuit, the center tap of T_2 detects this e_3 signal. Similarly, e_4 couples through as a common-mode signal on the e_{s3} wire pair to the center tap of T_4. Transformer T_5 extracts the phantom third or e_{s2} differential output signal by monitoring the voltage difference between the T_2 and T_4 center taps.

Thus with the phantom circuit, two pairs of wires transmit three differential signals without fundamental errors from ground-potential differences, noise pickup, or crosstalk. All three signal transmissions remain differential so the circuit automatically removes ground-potential differences and noise pick-up as described with Fig. 10.1b. In addition, the balanced impedances provided by transformer center tap connections and wire matching largely eliminate crosstalk. This balance typically reduces the crosstalk to −60 dB down from the level of the companion signals.[4] However, line voltage drops still reduce the transmission gains for all three differential signals, as described earlier with Fig. 10.3.

10.3   Implementing Phantom Amplifiers

As mentioned before, the classic phantom circuit encounters a fundamental limitation at lower frequencies where the size of the required transformers becomes impractical. Differential amplifiers replace the transformers to produce phantom amplifiers that extend the frequency response down to dc. Making this replacement requires attention to circuit grounding and a modification that adapts the phantom configuration to the differential amplifier equivalents of transformers. Careful grounding practices that accommodate these amplifiers avoid the reintroduction of the crosstalk errors described earlier. Modifying the phantom configuration adapts it to the differential amplifier's lack of a direct equivalent to a transformer center tap.

10.3.1   Grounding the phantom circuit

Including the common and ground connections in the evaluation of Fig. 10.6 permits an extension of the phantom circuit to more generalized instrumentation. At its inception, the phantom circuit coupled signals from floating, unpowered sources to loads that similarly required no ground reference. Developed for early telephone transmission, the phantom circuit originally coupled a microphone source directly to a speaker load with no external source of power. In more general applications, the source circuit conducts current from a power supply, like the case of a transducer bridge, and this requires a ground return path. Also, a common reference must be established for a monitoring system having multiple input sources and output loads.

Care must be taken in adding common and ground connections to the phantom circuit to avoid the reintroduction of errors from ground-potential differences and crosstalk. Figure 10.6 shows the correct grounding approach with a common reference connection for the input sources, a ground connection for the output signals, and a short-circuit connection between common and ground. On the circuit's input side, either input of a differential wire pair can be connected to common and the analogous condition holds for the ground connections of the output side. The separate return line that ties the common to the ground carries currents drawn from the system power supply to bias the source circuitry. This connection returns those currents without creating line voltage drops that could introduce the crosstalk described in Sec. 10.1.2.

This grounding convention also avoids errors from ground-potential differences. Normally in remote instrumentation, the output side of the system supplies power for the input side, making the output side the most reliable point for an earth ground. To avoid the error produced by ground-potential differences, the common reference on the input side does not directly connect to its own earth ground. Instead, this common only connects to that ground through a separate return line to the output side. This reduces the ground-potential difference to the voltage drop developed on that line, and the differential inputs of the circuit remain insensitive to this ground line drop.

10.3.2 Differential-amplifier equivalents to transformers

Transformers offer a simple implementation of the phantom circuit for most signal frequencies. However, lower-frequency signals require increasingly larger transformers to retain adequate signal coupling. Remote instrumentations that monitor physical parameters commonly encounter very low-frequency signals, making transformers impractical. For such cases, differential amplifiers provide transformer-like action extending down to dc. These amplifiers also remove much of the gain error and crosstalk previously introduced by line voltage drops.

To replace the transformers, these amplifiers transmit a differential input signal to a differential output with independent control of the accompanying common-mode signal. This amplifier action parallels the transformer operation described earlier with Figs. 10.4 and 10.5. Figure 10.7 demonstrates this parallel with the difference amplifier, an instrumentation amplifier, and the transformer equivalent. Note that all three produce the same differential and common-mode output signals from their e_1, e_2, and e_3 input signals. There, differential signal input e_s reproduces itself at the circuit's differential output but offset from ground by the e_3 signal drive.

While well recognized for their differential inputs, the difference amplifier and instrumentation amplifier generally serve single-ended out-

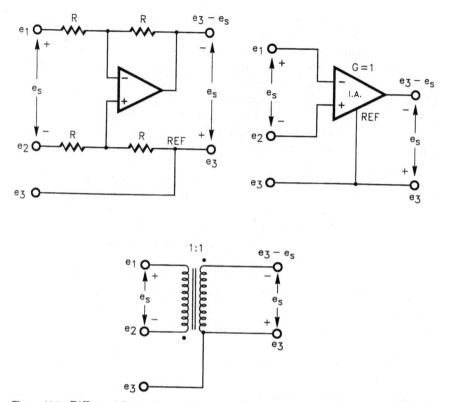

Figure 10.7 Differential amplifier action parallels that of a transformer but without the need for the impedance balancing of a center tap connection.

puts. However, as illustrated, both amplifiers present REF pins that serve as the reference return of the output signal. Normally, grounding this pin references the amplifier's output to ground for the single-ended output performance required in most applications. Returning this REF pin to a reference or control voltage shifts the output reference away from ground or even modulates that output reference level. Under these conditions, the amplifiers shown continue to supply their differential output voltages but referenced to the voltage at their REF pins. Thus like a transformer, these amplifiers provide a differential output free for common-mode control of their output signals.

The transformer equivalent to these amplifiers, also shown in Fig. 10.7, differs from the previous transformer connections in two respects. First, the transformer connection shown inverts the polarity of the differential signal in its input-to-output transmission. This change reproduces the phase inversion introduced by the differential amplifiers. The second transformer difference moves the output winding tap from the transformer's center to its bottom leg, paralleling the REF pin of the amplifiers. For the actual transformer case, this difference would be significant because it would insert a winding impedance between the e_3 signal and only one of the output lines. Previously, center tap drive of transformers assured equal winding impedances between e_3 and both transformer outputs to equalize the associated parasitic voltage drops. However, for the differential amplifier cases, the equivalent winding impedances reduce to zero, making this tap difference unimportant. The amplifiers shown accurately transmit e_3 to both lines of the transmission pair with no loss equivalent to that developed by a winding impedance.

10.3.3 Practical phantom amplifiers

Given the circuit equivalents of the preceding section, the substitution of differential amplifiers for transformers extends the low-frequency response of phantom circuits. The type of differential amplifier used for each transformer substitution depends on input impedance requirements. The difference amplifier offers the simplest substitution, but that amplifier presents a relatively low input resistance. That lower resistance remains acceptable on the input side of the phantom amplifier because the resulting input currents flow to the input common and its direct connection to the output ground. Given that connection, these input currents do not flow through the impedances of the differential transmission lines, where they could produce gain and crosstalk errors. However, on the output side of the circuit, the lower input resistance of the difference amplifier would draw significant input currents through those parasitic line impedances. This result would parallel the transformer case where transformers present low input impedances at the output side of the circuit. The higher input impedances of instrumentation amplifiers offer the option to virtually remove this line current and avoid the gain and crosstalk errors previously described. In addition, the low-level line currents of this solution permit the use of smaller-diameter wire for greater instrumentation economy.

To realize the benefits described, differential amplifiers replace the transformers of Fig. 10.6, as illustrated in Fig. 10.8. In this replacement, differential amplifiers A_1 through A_5 substitute for the similarly numbered transformers of the earlier figure. For comparison simplicity, this figure illustrates the differential amplifiers with unity voltage gains but the circuit remains capable of higher gains in the output amplifiers. In the figure, the difference amplifier configuration replaces the previous T_1 and T_3 input transformers for the greatest economy, since this input application tolerates lower input resistances. At the output, high-input-impedance instrumentation amplifiers replace the

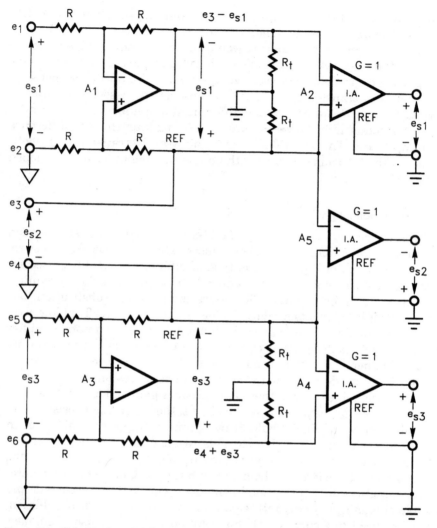

Figure 10.8 Substitution of differential amplifiers for transformers converts Fig. 10.6 to a phantom amplifier with a frequency response extending down to dc.

T_2, T_4, and T_5 transformers. The circuit also shows four R_t input return resistors that provide local ground paths for the instrumentation amplifier input currents. These resistors do shunt the input impedances of the instrumentation amplifiers, so higher resistances should be used there to avoid the reintroduction of line voltage drops.

With this configuration, the circuit operation parallels that of the preceding transformer-coupled phantom circuit. Differential input signals e_{s1} and e_{s3} initially transfer to the outputs of A_1 and A_3, having any accompanying common-mode signals removed by those amplifiers. Once free of those common-mode signals, the introduction of the e_3 and e_4 drive of the A_1 and A_3 REF pins adds intentional common-mode components to the transmission of e_{s1} and e_{s3}. The combined differential and common-mode signals transfer through the two wire pairs to the output amplifiers. At that point, output amplifiers A_2 and A_4 reject the common-mode components of e_{s1} and e_{s3} to present those signals as ground-referenced outputs. The third or phantom signal e_{s2} transmits as the difference between the e_3 and e_4 common-mode signals intentionally introduced. Amplifier A_5 detects this difference and supplies $e_{s2} = e_4 - e_3$ as a third, ground-referenced output signal. In this process, spurious coupling also introduces common-mode signals that could interfere, but the differential detection of A_5 rejects those effects.

10.3.4 Error reduction with phantom amplifiers

Common-mode characteristics of the amplifiers of Fig 10.8 determine the circuit's actual rejection of crosstalk, ground-potential differences, and noise pickup. Each of these error sources reacts with the amplifier common-mode rejection ratios to develop residual differential error signals. Crosstalk no longer results from line drops, due to the wiring isolation retained here from earlier discussions. However, the introduced common-mode signals e_3 and e_4 develop differential errors for the two pairs of wires, and these errors represent crosstalk components. Signal e_3 first exercises the common-mode input swing of A_1 and adds an error signal of e_3/CMRR_1 to the differential output of that amplifier. Next, e_3 reacts with the common-mode rejection of A_2 to develop an additional differential error of e_3/CMRR_2.

Together, these common-mode error signals produce an output crosstalk error for the e_{s1} channel of $e_3(1/\text{CMRR}_1 + 1/\text{CMRR}_2)$ at the A_2 output. An analogous crosstalk signal of $e_4(1/\text{CMRR}_3 + 1/\text{CMRR}_4)$ develops at the output of A_4 for the e_{s3} channel. Still, the net crosstalk decreases to a far lower level with this amplifier version of the phantom circuit. In part, this improvement results from the very high common-mode rejection ratios of the amplifiers which far exceed the accuracy of a transformer's center tap connection. In addition, the low input currents of the amplifiers greatly reduce crosstalk sensitivity to line impedance imbalances.

Other common-mode signals, created by ground-potential differences and noise pickup, receive similar common-mode rejection. However, these error signals potentially impose a new range limit in the differential amplifier case, since these amplifiers restrict the common-mode signal range more than the transformer original. Transformers readily accept hundreds of volts of common-mode signals while the signals permissible with the differential amplifiers often remain less than those of the power supplies that bias them. This reduced common-mode range seldom produces a restriction to noise rejection but can easily be exceeded by ground-potential differences. Generally, adding the direct connection between the input common and output ground, as illustrated, restricts ground-potential differences to levels within the amplifiers' common-mode range. However, for this reason, the input circuit should remain floating with respect to earth ground potential except for its return line to the circuit's output ground.

References

1. J. Graeme, "Creating Phantom Circuits Simplifies Remote Monitoring," *EDN*, p. 123, August 1991.
2. H. Ott, *Noise Reduction Techniques in Electronic Systems*, 2d ed., Wiley, New York, 1988.
3. D. Tucker, "A Technical History of Phantom Circuits," *Proc. IEE*, vol. 126, no. 9, September 1979.
4. Tucker, op cit., 1979.

Glossary

bootstrap Circuit technique that removes undesired effects of a source by driving the source's normal common return with the signal delivered by the source itself.

bridge amplifier Amplifier that translates the differential output signal of a transducer bridge into an amplified, ground-referenced output signal.

clamping amplifier Amplifier that produces a linear output in response to an input signal within a predetermined range and clamps its output voltage at a fixed level outside that range.

common-mode feedback Feedback control of a common-mode rather than differential signal, such as in the removal of a common-mode signal from a differential output.

common-mode input capacitance C_{icm} Effective capacitance between either input of a differential amplifier and common ground.

common-mode rejection (CMR) Logarithmic form of common-mode rejection ratio as expressed by CMR = 20 log(CMRR).

common-mode rejection ratio (CMRR) Ratio of the differential gain of an amplifier to its common-mode gain A_D/A_{CM}.

common-mode voltage Average of the voltages appearing on the two lines of a differential transmission, such as those at the inputs or outputs of differential amplifiers.

composite amplifier Op amp circuit enclosing two or more op amps within a common feedback loop.

difference amplifier Op amp circuit having a combination of inverting and noninverting voltage gains that produces the response of a differential-input amplifier.

differential-input amplifier   Amplifier that transmits a signal proportional to the difference between two input signals while rejecting their common-mode component.

differential-input capacitance C_{id}   Effective capacitance between the two inputs of a differential amplifier.

differential noninverting amplifier   Combination of two noninverting op amp circuits that produces an amplified differential output signal in response to a differential input signal.

feedback   Return of a portion of the output signal of a device to its input for response control.

feedback factor ß   In a feedback system, fraction of the output fed back to the input.

feedback intercept   *See* intercept frequency.

frequency compensation   *See* phase compensation.

gain error   For an op amp with feedback, difference between the actual closed-loop gain and that predicted by the ideal gain expression.

gain error signal   Differential signal voltage e_o/A developed by feedback between the two inputs of an op amp as a result of the amplifier's finite open-loop gain.

gain margin   Gain magnitude difference separating an application circuit's 1/ß intercept from that at which the feedback loop's phase shift reaches 360°.

input bias current I_B   Dc biasing current drawn by the inputs of an op amp.

input capacitance   *See* common-mode input capacitance, differential input capacitance.

input error signal e_{id}   Combined differential signal voltage, developed by feedback, between the two inputs of an op amp as a result of the amplifier's error sources.

input offset current I_{OS} Difference between the two input bias currents of an op amp.

input offset voltage V_{OS} Dc voltage between the two inputs of an op amp that drives its output voltage to zero.

instrumentation amplifier Differential-input, differential-output amplifier with internal feedback committed for voltage gain.

intercept frequency f_i Frequency at which an op amp circuit's $1/\beta$ response intercepts the amplifier's A_{OL} response, marking the highest frequency for which the amplifier's available gain supplies the feedback gain demand.

inverting amplifier Amplifier with a negative gain magnitude that inverts signal polarity.

loop gain $A\beta$ Excess gain available to supply increasing feedback demand as represented by the separation between an op amp's open-loop response and a given circuit's $1/\beta$ response.

minimum-stable gain Minimum closed-loop gain at which an op amp can be operated with frequency stability.

noise gain Gain that amplifies the input noise voltage of an op amp and follows the $1/\beta$ response of an application circuit up to the intercept frequency.

noninverting amplifier Amplifier with a positive gain magnitude that does not invert signal polarity.

offset current *See* input offset current.

offset voltage *See* input offset voltage.

open-loop gain A Ratio of an op amp's output signal magnitude to the associated signal magnitude appearing between that amplifier's two inputs.

phantom amplifier Amplifier equivalent of the phantom circuit that replaces transformers with differential amplifiers for extended low-frequency response.

phantom circuit Transformer-coupled circuit that transmits three differential signals for every two two-wire pairs as though a third or phantom wire pair were present.

phase compensation Frequency response tailoring for stabilizing a feedback system through the addition of response poles and zeros that reduce high-frequency phase shift.

phase margin ϕ_m Margin separating the phase shift around a feedback loop from $360°$ at the unity loop-gain point of the intercept frequency f_i.

power-supply rejection (PSR) Logarithmic form of power-supply rejection ratio as expressed by PSR = 20 log(PSRR).

power-supply rejection ratio (PSRR) For an op amp, ratio of a power-supply voltage change to the resulting differential input error e_{id}.

rate of closure Difference in slopes of the $1/ß$ and A_{OL} responses at their crossing as expressed in dB.

standard denominator, 1 + 1/Aß Response denominator that transfers the common gain and stability analysis results to all op amp feedback connections.

summing amplifier Op amp circuit in which multiple input signals couple through scaled resistances to develop an output signal with weighted dependencies on the various input signals.

summing junction For an inverting amplifier, junction of an op amp's feedback elements that drives the amplifier's input and permits signal summation through feedback reduction of that junction's voltage to zero.

tee network Arrangement of passive elements in the shape of a T, commonly used for op amp feedback in realizing the equivalents of impedance extremes or in response tailoring.

transconductance amplifier Amplifier that converts an input signal voltage to an output signal current through a defined transconductance gain.

transimpedance amplifier   Amplifier that converts an input signal current to an output signal voltage through a defined transimpedance gain.

unity-gain crossover f_c   Frequency at which the open-loop gain of an op amp crosses unity, or 0 dB.

virtual ground   Groundlike characteristic of the inverting input of an op amp where feedback absorbs injected current without developing a voltage.

voltage follower   Short-circuit feedback connection of an op amp which results in an output signal that follows the signal applied at the amplifier's noninverting input.

Index

ABOUT THE AUTHOR

Jerald G. Graeme is principal engineer at Gain Technology in Tucson, Arizona. He has 30 years of design, management, and training experience in linear IC product development and holds 9 U.S. patents. He has written 5 other best-selling circuit design books for McGraw-Hill and more than 100 articles for publications including *EDN* and *Electronic Design*. He hods a BSEE degree from the University of Arizona and an MSEE degree from Stanford University.

SOFTWARE AND INFORMATION LICENSE

The software and information on this diskette (collectively referred to as the "Product") are th property of The McGraw-Hill Companies, Inc. ("McGraw-Hill") and are protected by both Unite States copyright law and international copyright treaty provision. You must treat this Product jus like a book, except that you may copy it into a computer to be used and you may make archiva copies of the Products for the sole purpose of backing up our software and protecting your invest ment from loss.

By saying "just like a book," McGraw-Hill means, for example, that the Product may be used b any number of people and may be freely moved from one computer location to another, so long a there is no possibility of the Product (or any part of the Product) being used at one location or o one computer while it is being used at another. Just as a book cannot be read by two different peo ple in two different places at the same time, neither can the Product be used by two different peo ple in two different places at the same time (unless, of course, McGraw-Hill's rights are being violated).

McGraw-Hill reserves the right to alter or modify the contents of the Product at any time.

This agreement is effective until terminated. The Agreement will terminate automatically with out notice if you fail to comply with any provisions of this Agreement. In the event of termination by reason of your breach, you will destroy or erase all copies of the Product installed on any com puter system or made for backup purposes and shall expunge the Product from your data storage facilities.

LIMITED WARRANTY

McGraw-Hill warrants the physical diskette(s) enclosed herein to be free of defects in materials and workmanship for a period of sixty days from the purchase date. If McGraw-Hill receives writ ten notification within the warranty period of defects in materials or workmanship, and such noti fication is determined by McGraw-Hill to be correct, McGraw-Hill will replace the defective diskette(s). Send request to:

Customer Service
McGraw-Hill
Gahanna Industrial Park
860 Taylor Station Road
Blacklick, OH 43004-9615

The entire and exclusive liability and remedy for breach of this Limited Warranty shall be limited to replacement of defective diskette(s) and shall not include or extend to any claim for or right to cover any other damages, including but not limited to, loss of profit, data, or use of the software, or spe cial, incidental, or consequential damages or other similar claims, even if McGraw-Hill has been specifically advised as to the possibility of such damages. In no event will McGraw-Hill's liability for any damages to you or any other person ever exceed the lower of suggested list price or actual price paid for the license to use the Product, regardless of any form of the claim.

THE McGRAW-HILL COMPANIES, INC. SPECIFICALLY DISCLAIMS ALL OTHER WAR-RANTIES, EXPRESS OR IMPLIED, INCLUDING BUT NOT LIMITED TO, ANY IMPLIED WARRANTY OF MERCHANTABILITY OR FITNESS FOR A PARTICULAR PURPOSE Specifically, McGraw-Hill makes no representation or warranty that the Product is fit for any par ticular purpose and any implied warranty of merchantability is limited to the sixty day duration of the Limited Warranty covering the physical diskette(s) only (and not the software or in-formation) and is otherwise expressly and specifically disclaimed.

This Limited Warranty gives you specific legal rights; you may have others which may vary from state to state. Some states do not allow the exclusion of incidental or consequential damages, or the limitation on how long an implied warranty lasts, so some of the above may not apply to you.

This Agreement constitutes the entire agreement between the parties relating to use of the Product. The terms of any purchase order shall have no effect on the terms of this Agreement. Failure of McGraw-Hill to insist at any time on strict compliance with this Agreement shall not constitute a waiver of any rights under this Agreement. This Agreement shall be construed and governed in accordance with the laws of New York. If any provision of this Agreement is held to be contrary to law, that provision will be enforced to the maximum extent permissible and the remaining provisions will remain in force and effect.